KT-196-175

THE JAPANESE ABACUS EXPLAINED

by

Y. YOSHINO

with an introduction by
MARTIN GARDNER

Dover Publications, Inc.

New York

Published in Canada by General Publishing Company, Ltd., 30 Lesmill Road, Don Mills, Toronto, Ontario.

Published in the United Kingdom by Constable and Company, Ltd., 10 Orange Street, London W. C. 2.

This Dover edition, first published in 1963, is an unabridged and unaltered republication of the work first published by Kyo Bun Kwan, Tokyo, in 1937.

This volume also contains a new Introduction by Martin Gardner especially prepared for this Dover edition.

International Standard Book Number: 0-486-21109-6
Library of Congress Catalog Card Number: 63-20256

Manufactured in the United States of America

Dover Publications, Inc.
180 Varick Street
New York 14, N. Y.

Introduction to Dover Edition

Why should anyone in the western world, in this day of versatile office calculating machines and mammoth electronic computers, waste his time learning how to use a simple abacus? There are several reasons.

In the first place, the office calculators are expensive. Who wants to spend a hundred dollars or more on a computer to use, outside a business office, on those rare occasions when a long list of numbers has to be added? The abacus, in contrast, is ideal for home use or by a small shopkeeper. It costs little, operates reliably, requires no maintenance, can be carried in the coat pocket, and makes an unusual pattern hanging on a wall.

In the second place, using an abacus is *more fun* than adding or subtracting on paper. For many years, ever since I was introduced to the abacus by a friend, L. Vosburgh Lyons, I have used it for adding figures in check books, totaling store receipts, and on countless other occasions when a tedious numerical calculation presented itself. The tactile sensation of clicking the beads back and forth turns a dull task into a pleasant one. The mind has less work to do than when figuring with pencil and paper, and there is a pleasure in developing the finger skills that make for rapid manipulation of the beads.

In the third place, working with the abacus introduces one to an interesting segment of eastern culture. Not only is the abacus still widely used throughout Japan, China and India, but in the Soviet Union as well. If you buy more than one item in any Moscow store the chances are ten to one that the storekeeper will tally up the prices on an abacus. Learning to use the device is something like developing a taste for Japanese *sukiyaki* or Russian *kvas*. It helps in its small way to bridge cultural gaps. Lyons once told me that he was in a Chinese laundry one day where he overheard a surly gentleman demanding to know how much he owed. "Hurry up and figure it out," the man said, "on that crazy thing there of yours." The remark is typical of a barbarian in a minor state of what anthropologists call "cultural shock." Anything alien to his own experience is inferior, childish, a bit "crazy."

Finally, there is (at least for me) a pleasure in rebelling against the enormous overcomplexity of modern life. "Simplicity, simplicity, simplicity!" cried Henry David Thoreau. "I say, let your affairs be as two or three, and not a hundred or a thousand; instead of a million count half a dozen, and keep your accounts on your thumb nail." The abacus could hardly be simpler; yet it does exactly what it is supposed to do, with great efficiency, without electrical switches, vacuum tubes, transistors; without even a single lever or cogwheel. Discovering it is like a modern housewife's discovery that an excellent clothes dryer can be made from a piece of rope stretched horizontally. On the wall, near a giant computer, one sometimes sees an abacus behind glass, with the sign: IN CASE OF EMERGENCY, USE THIS. I for one take a perverse delight in operating a digital computer so simple that even the ancient Romans were able to manufacture it.

A skilled abacus operator can hold his own with a skilled machine operator, except on large problems of multiplication and division. Whenever a contest in speed and accuracy is staged between an abacus and a modern desk calculator, the abacus usually wins. In 1946, wide publicity was given to a contest in Tokyo (on November 11 in the Ernie Pyle Theater) between Private Thomas Nathan Wood, of Deering, Missouri, using an American office machine, and Kiyoshi Matsuzaki, a local abacus expert. Private Wood was not a skilled operator, so perhaps the contest was a bit rigged*; nevertheless, the audience was astounded when Matsuzaki won. A week later a similar contest was staged in New York for broadcast over radio station WOR. A huge office machine was wheeled into the studio on a typewriter table, for operation by Dorothy Boudreau of the WOR payroll department. Mr. P. T. So, a Chinese student at Columbia University, took his wooden abacus out of his brief case. Mr. So, who later apologized for being so clumsy, finished eight seconds ahead of his opponent.

The Chinese abacus or *suan-pan,* by the way, is a bit different from the Japanese *soroban.* It has an extra row of beads on top, another extra row on the bottom. Both rows are little used in ordinary computation, and since the nineteenth century have been dropped entirely from the modern *soroban.* However, if you happen to own a Chinese abacus instead of a Japanese model, there is no need to discard it. Everything in this clearly written, elementary manual is applicable; just ignore the top and bottom rows of beads.

For practice in addition, let me recommend the following amusing exercise. Place on the abacus the number 12345679 (note omission of the 8). Add this number to itself eight times, which is the same as multiplying it by 9. The result will be a sum consisting entirely of 1's. Add 12345679 nine more times to get a row of 2's. Nine more times produces a row of 3's, and so on until you have added the number 80 times to finish with 999,999,999. It is a good exercise because it exploits every variety of bead manipulation, and in a series of stages that are easy to check. By timing yourself each time you com-

* See the letter by Major A. M. Maish in *Mathematics Magazine,* Vol. 29, No. 1, September-October, 1955, giving details of the contest not covered in the wire service accounts.

plete a stage correctly you can see how your speed is improving. An expert can do each stage of nine additions in half a minute.

Happy bead flicking!

Dobbs Ferry, N. Y.
July, 1963

MARTIN GARDNER

FOREWORD

The following pages have been written for High School and College students, either boys or girls, and for anyone who may be interested in a study of the abacus.

The author, when he voyaged to the United States some years ago, found many friends there who wanted to know about the abacus or Japanese *soroban*. It was understood that the abacus had educational value, not to mention its usefulness, but they could not find an account of it in any of the textbooks in use.

In the pages that follow, methods of calculation on the abacus, in use in Japan from ancient times to the present, are described.

The abacus is more accurate and a quicker instrument of calculation than the use of figures written down. In size and cost of production, it is more convenient and less expensive than calculating machines.

For this reason, the abacus is recognized in commercial circles in Japan as the best instrument for making all calculations.

This book has been written in simple and plain English, with numerous illustrations or diagrams, in order to make the subject as clear as possible to the reader. It should be borne in mind that the abacus is used with almost incredible speed in making calculations. It is also not to be overlooked that a mental process accompanies the manipulation of the counters with the fingers.

The author hopes that all his American friends will be helped, by the use of this treatise, to understand and use the abacus, and that they may be able to acquire the speed and accuracy achieved in every day life here in Japan. If so, the author's purpose in writing this book will have been fulfilled.

Y. Yoshino, author
Kuragano, Japan
February 15, 1937

CONTENTS

Page

Chapter I Abacus 1

Section 1. The names of the parts of an abacus
Section 2. The value of the counters
Section 3. How to place and to return the counters
Section 4. The proper use of the fingers

Chapter II Addition 6

Section 1. Addition table
Section 2. Practical use of the counters in addition
Part 1. Adding with numbers of one place
Part 2. The determination of the position on a board
Part 3. Illustration of large number on a board
Part 4. Adding with large numbers
Section 3. Addition from dictation
Section 4. Addition by oneself
Section 5. Addition of figures from tabulated forms

Chapter III Subtraction 54

Section 1. Subtraction table

Section 2. The practical use of the counters in subtraction

Part 1. Subtracting with numbers of one place

Part 2. Subtracting with large numbers

Section 3. Subtraction in cases where the subtrahend is larger than the minuend

Chapter IV Multiplication 89

Section 1. Multiplication table

Section 2. Determination of the position in multiplication

Section 3. Multiplication beginning at the right

Part 1. The process of multiplication

Part 2. Examples

Section 4. Multiplication beginning at the left

Part 1. The process of multiplication

Part 2. The determination of the unit position

Part 3. Examples

Section 5. Multiplying with decimals

Section 6. Multiplying tenths and hundredths

Chapter V Division 155

Section 1. Division by unification
Part 2. Examples
Part 3. Division table by transfiguration
Part 4. Practical use of the division table by transfiguration
Part 5. Division table by replacement
Part 6. Practical use of division table by replacement
Section 2. The determination of position
Section 3. The process of division
Part 1. Practical training in division
Part 2. Exercises

Chapter VI Division by abbreviation 222

Section 1. The determination of position
Section 2. The process of division by abbreviation type
Section 3. The process of division by abbreviation type
Section 4. Practical training in division by abbreviation

For the Salesman 238

CHAPTER I.

ABACUS

Section 1.

The names of the parts of an abacus.

A is the crosspiece
B is an upright
C is the upper part or Heaven
D is the lower part or Earth

The crosspiece divides the abacus into two compartments. The upper one is called "Heaven" and the lower one "Earth."

An upright serves as an axis for the counters.

The counters in "Heaven" are generally worth five points while those in "Earth" are worth only one. Before beginning to use the abacus all the Heaven counters must be pushed up to the top of their uprights. This can be done with one sweeping movement of the fingers.

The uprights on the left hand side form the "upper position" and those on the right the "lower position."

For example, if we wish to show the number 152 on the abacus we must begin from the left as is shown in the following illustration.

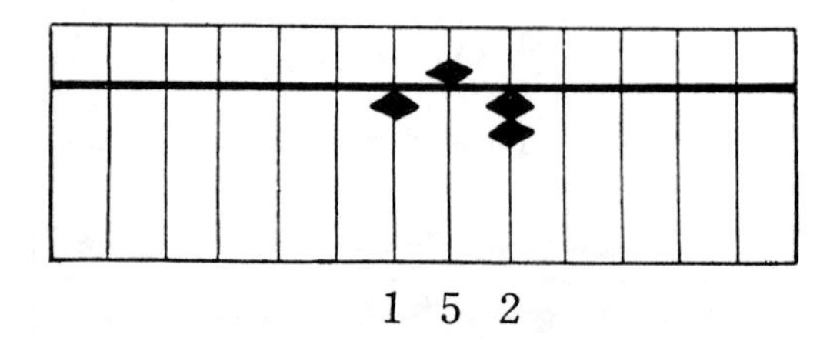

In this case the first upright stands for one hundreds, the second for tens, and the third for ones. Therefore we need one one-hundreds, five tens, and two ones.

Section 2.

The value of the counters.

A counter in Heaven is worth 5.

A counter in Earth is worth 1.

Therefore one counter in Heaven is worth five counters in Earth and is generally spoken of as the 5 counter.

Section 3.

How to place and to return the counters.

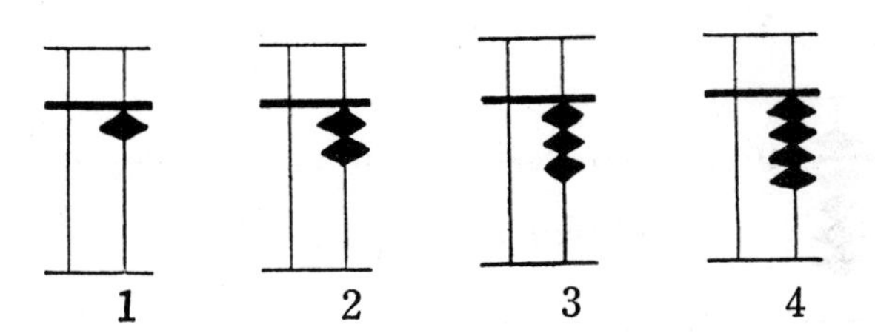

To designate the numbers 1, 2, 3, and 4 the Earth counters are used.

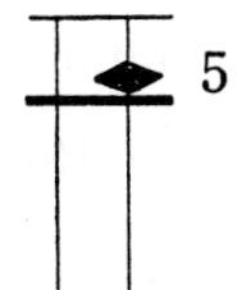

To designate 5 a counter of Heaven is used.

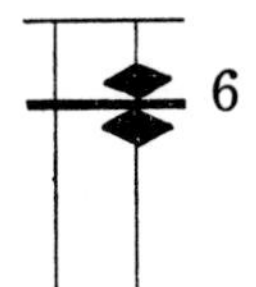

To designate 6 a counter of Heaven and one of Earth are used.

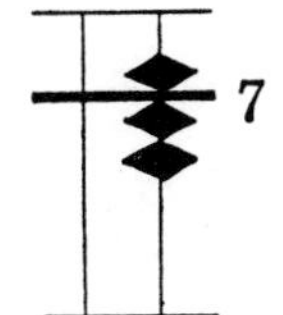

To designate 7 a counter of Heaven and two of Earth are used.

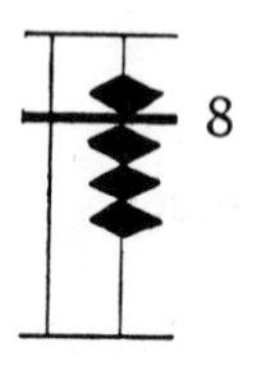

To designate 8 one counter of Heaven and three of Earth are used.

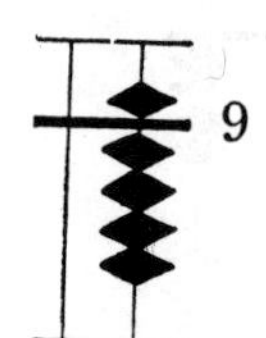

To designate 9 one counter of Heaven and four of Earth are used.

As the above illustrations show when we use the counters for numbers less than 5 we lift into place the Earth counters beginning with the top one. When the number is larger than 5 we push down into place one of the Heaven counters and then lift up the Earth counters as needed.

As for example the following illustration shows the numbers 5, 7, and 12 in position.

5 and 7 are 12.

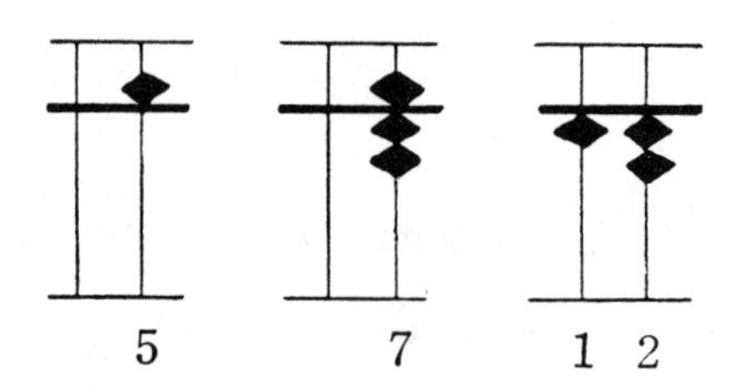

If the number 5 is indicated on the abacus and we wish to add to it the number 7, we must raise into position two Earth counters on the same upright as the 5 and at the same time push back into place the 5 counter and raise one Earth counter on the next upright to the left. Thus we see that the sum is 12.

The uprights on the left are always of higher value than those on the right so if we raise one earth counter on each of three uprights in succession from left to right we get the number one hundred and twenty three.

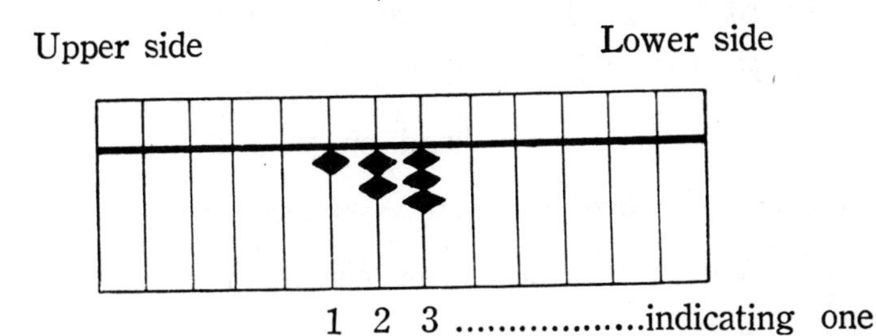

1 2 3 ....indicating one hundred and twenty three.

Section 4.

The Proper Use of the Fingers.

In handling the counters we should use two fingers, the thumb and forefinger.

The forefinger should be used to place into position the 5 counters and to return to place the Earth counters after using, and sometimes to return to place the Heaven counters.

The thumb should be used to raise into position the Earth counters and sometimes to return to place the 5 counters.

Care should be taken to train the thumb and the forefinger to perform these duties.

I should like to mention some other things about using the abacus.

(1) Keep a good posture, bending forward a little.

(2) Hold the back of the abacus in your left palm and do not lay your right hand on your desk as it must be kept free to move any counter on the board.

(3) When using the abacus clinch your right hand together with the exception of the thumb and forefinger.

(4) Do not spread your fingers.

(5) You must train yourself to recognize the figures as quickly as possible and accurately.

(6) You must recognize figures instantly and in proportion to your progress learn to recognize two or three figures at a time.

CHAPTER II.

ADDITION

Addition is the summing up of two or more figures.

Section 1.

Addition Table.

Memorize the following table.

$1+9=10$	One	and	Nine	are	Ten.
$2+8=10$	Two	and	Eight	are	Ten.
$3+7=10$	Three	and	Seven	are	Ten.
$4+6=10$	Four	and	Six	are	Ten.
$5+5=10$	Five	and	Five	are	Ten.
$6+4=10$	Six	and	Four	are	Ten.
$7+3=10$	Seven	and	Three	are	Ten.
$8+2=10$	Eight	and	Two	are	Ten.
$9+1=10$	Nine	and	One	are	Ten.

Section 2.

Practical use of the counters in addition.

From the explanations in the preceding chapters addition should be clear by now but as it is necessary to understand well in the beginning let me show you a practical addition in the following problem.

Find the sum of 5 and 1.

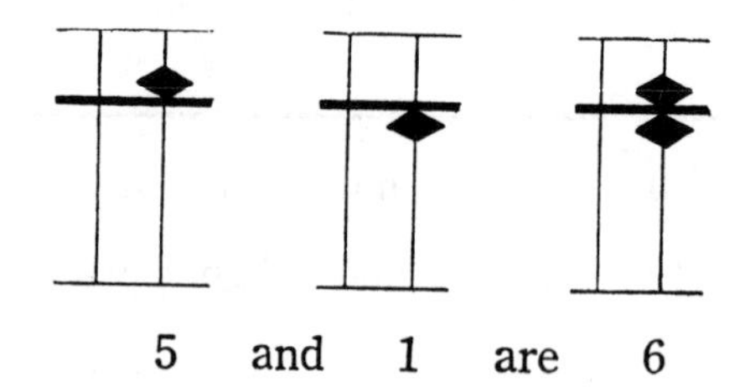

The use of the fingers:

With the forefinger, slide down the 5 counter first.

With the thumb, lift up the 1 counter on the same upright.

The sum is indicated as 6.

A few examples will give you practical training which will help you to understand. In these exercises the abbreviation "A. T." means "Addition Table."

Part One.

Adding with numbers of one place.

Example 1. 5 and 1 are 6.

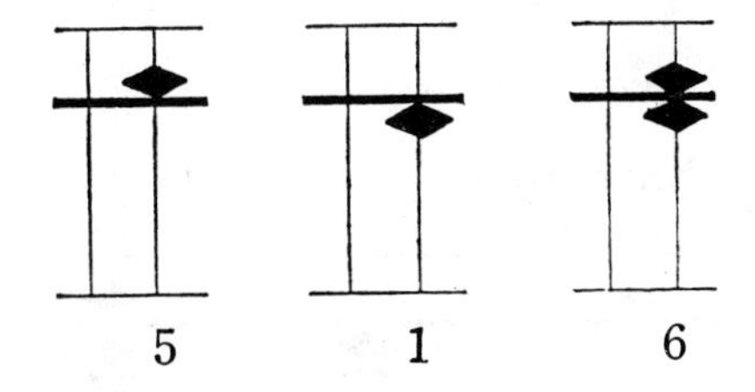

(1) Slide down a 5 counter with the forefinger.

(2) Raise a 1 counter on the same upright with the thumb.

The sum 6 is indicated on the board.

Example 2. 5 and 2 are 7.

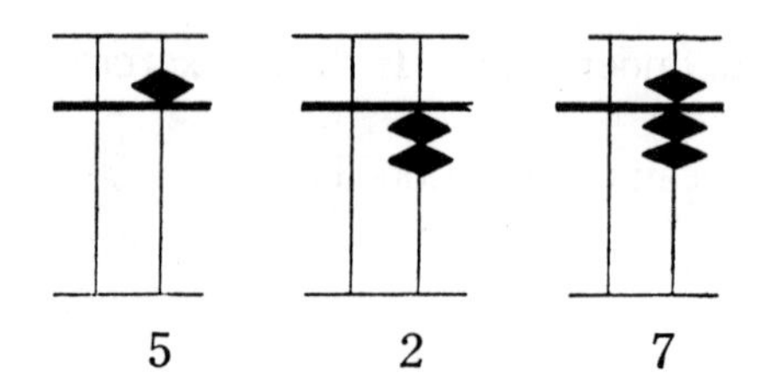

(1) Slide down a 5 counter with the forefinger.

(2) Raise two of the Earth counters on the same upright, with the thumb.

The sum 7 is indicated on the board.

Example 3. 5 and 3 are 8.

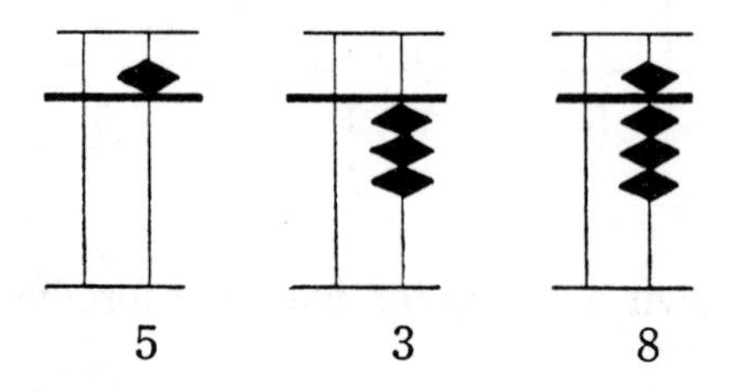

(1) Slide down a 5 counter with the forefinger.

(2) Raise three of the earth counters on the same upright, with the thumb.

The sum 8 is indicated on the board.

Example 4. 5 and 4 are 9.

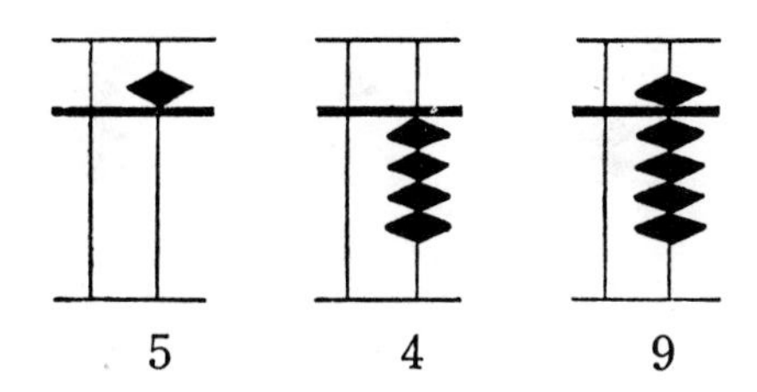

(1) Slide down one of the Heaven counters with the forefinger.

(2) Raise four of the Earth counters which are on the same upright as the Heaven counter, using the thumb.

The sum 9 is indicated on the board.

Example 5. 1 and 5 are 6.

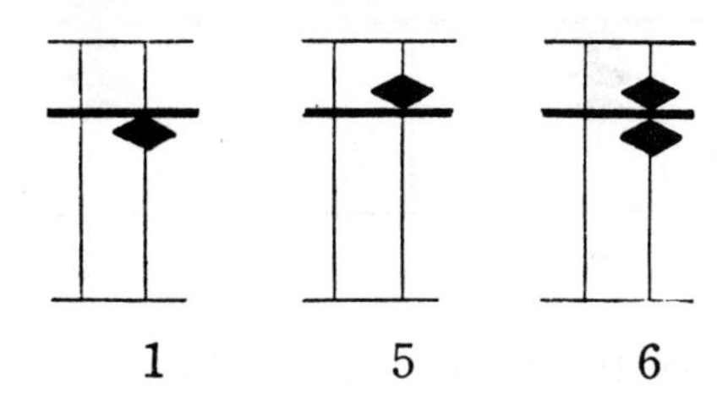

(1) Raise one Earth counter with the thumb.

(2) Slide down a Heaven counter on the same upright, with the forefinger.

The sum 6 is indicated on the board.

Example 6. 2 and 5 are 7.

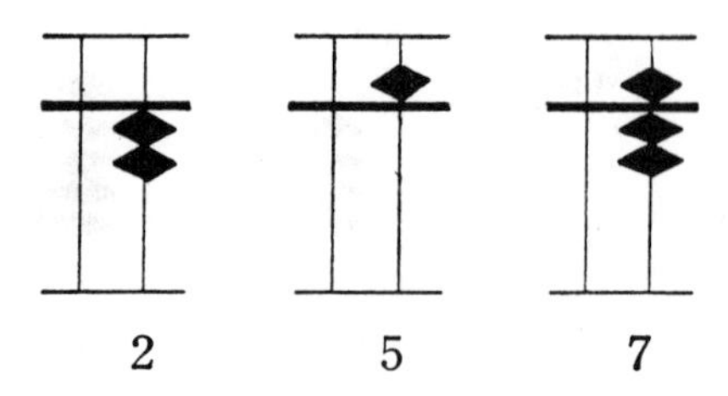

(1) With the thumb raise two Earth counters.

(2) With the forefinger slide down one Heaven counter on the same upright as the two 1 counters.

The sum 7 is indicated on the board.

Example 7. 3 and 5 are 8.

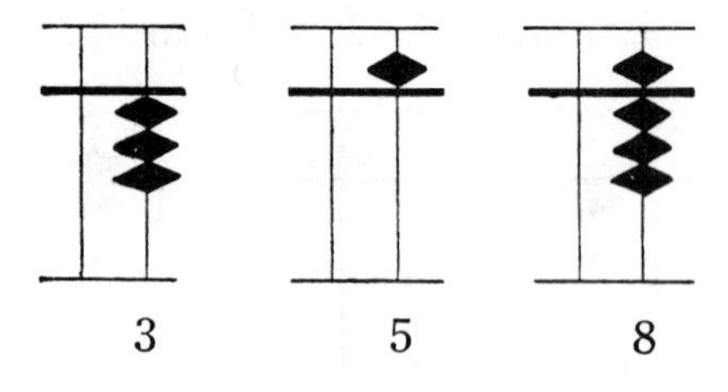

(1) With the thumb lift into place three of the 1 counters.

(2) With the forefinger slide down one of the 5 counters on the same upright as the three 1 counters.

The sum 8 is indicated on the board.

Example 8. 4 and 5 are 9.

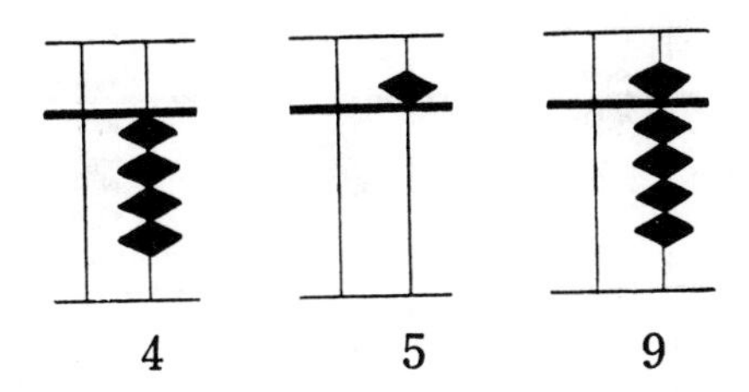

(1) With the thumb raise into position four of the one counters.

(2) With the forefinger slide down into position the 5 counter on the same upright as the four one counters.

The sum 9 is indicated on the board.

Example 9. 6 and 1 are 7.

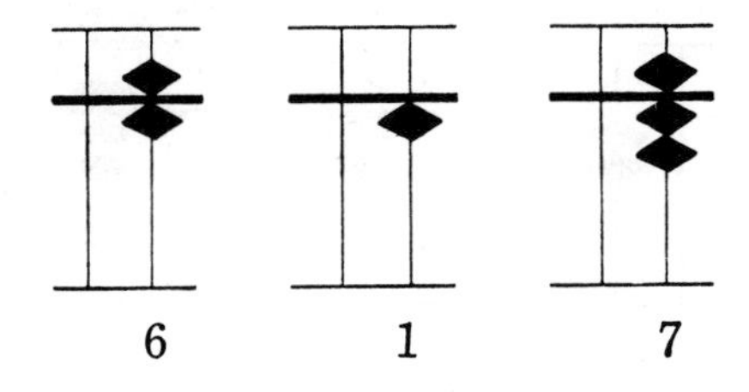

(1) With one movement press down a 5 counter and slide up a one counter, the forefinger on the Heaven counter and the thumb on the Earth one.

(2) With the thumb raise one Earth counter on the same upright with the other two. The sum 7 is indicated on the board.

Example 10. 6 and 3 are 9.

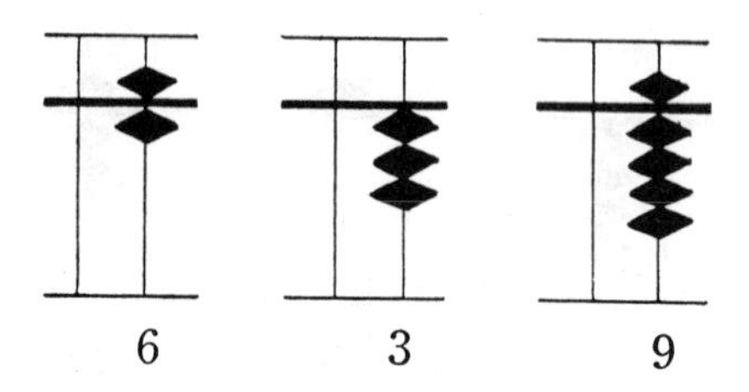

(1) As before with one movement press down a 5 counter and shove up a 1 counter with the forefinger and the thumb.

(2) With the thumb raise three of the one counters on the same upright with the other two.

The sum 9 is indicated on the board.

Example 11. 2 and 7 are 9.

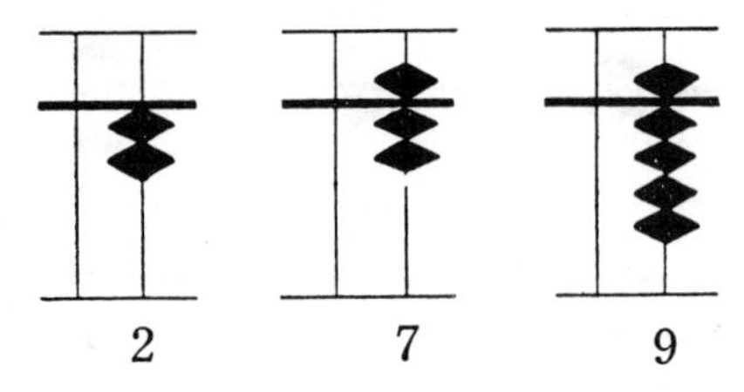

(1) With the thumb raise two Earth counters.

(2) With one movement raise two Earth counters with the thumb and one Heaven counter with the forefinger, on the same upright with the two Earth counters. Thus one Heaven counter and four Earth counters in position indicate the sum 9.

Example 12. 3 and 6 are 9.

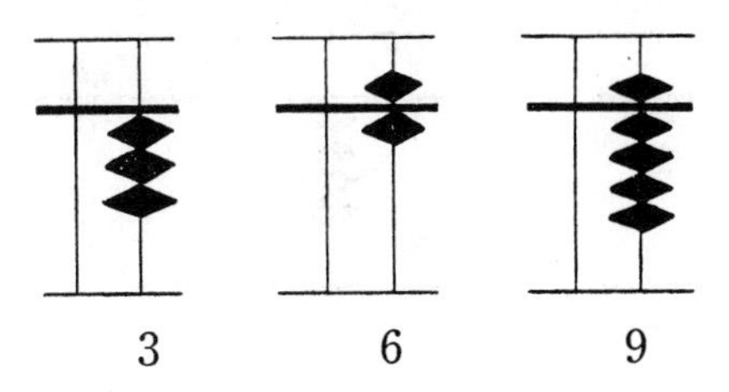

(1) Raise three Earth counters with the thumb.

(2) With one movement push down one Heaven counter with the forefinger and push up one Earth counter on the same upright with the thumb.

9 is indicated on the board. This is the sum.

Example 13. 1 and 4 are 5.

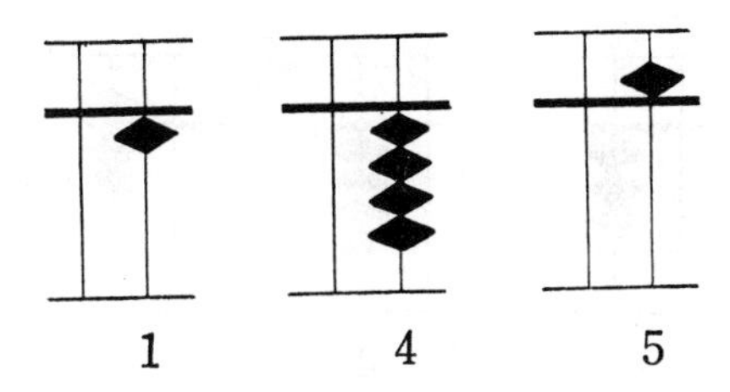

(1) With the thumb raise one Earth counter.

(2) Think "1 and 4 are 5." With the forefinger take down one of the five counters on the same upright as the one Earth counter and at the same time replace the one counter which was moved into position.

The sum 5 is indicated on the board.

Example 14. 2 and 4 are 6.

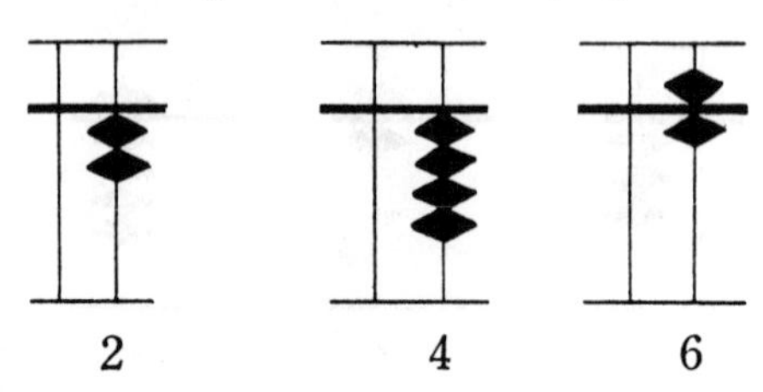

(1) With the thumb raise two Earth counters.

(2) Think, "4 and 1 are 5." With the forefinger slide down the Heaven counter on the same upright as the two Earth counters and return one of the Earth counters to its original position.

The sum 6 is indicated on the board.

Example 15. 3 and 4 are 7.

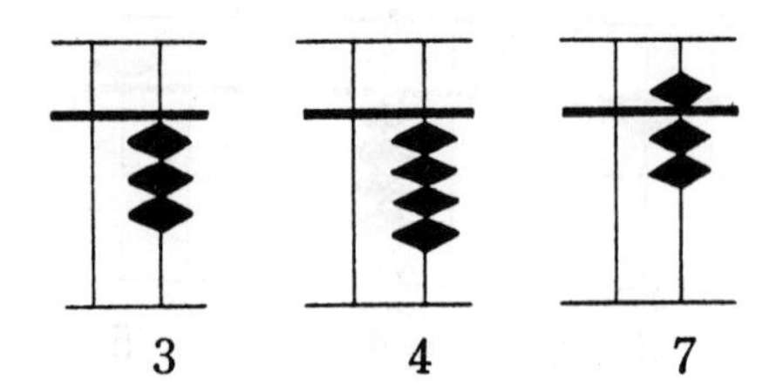

(1) With the thumb raise three Earth counters.

(2) Think, "4 and 1 are 5." Take down the Heaven counter on the same upright as the three Earth counters and at the same time with the same forefinger replace one of the Earth counters.

The sum 7 is indicated on the board.

Example 16. 4 and 4 are 8.

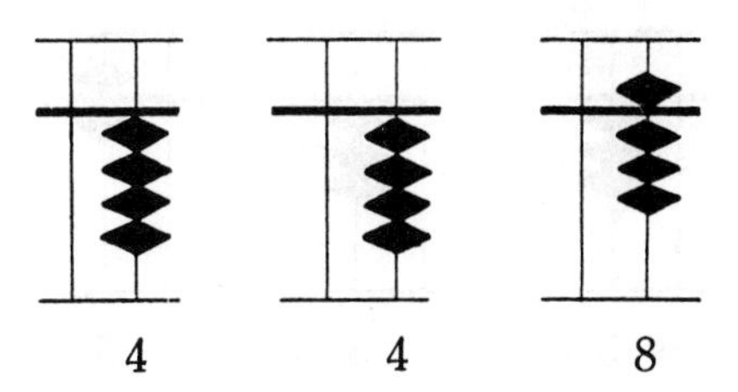

(1) With the thumb raise four of the one counters.

(2) Think, "4 and 1 are 5." Slide down the Heaven counter on the same upright with the four Earth counters and at the same time replace one of the Earth counters with the forefinger.

The sum 8 is indicated on the board.

Example 17. 4 and 3 are 7.

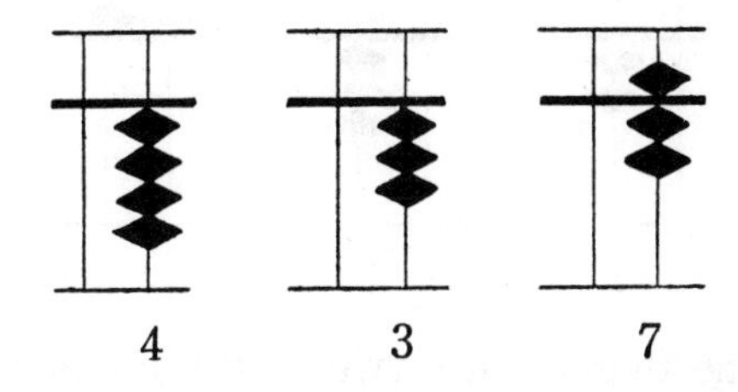

(1) With the thumb raise four Earth counters.

(2) Think, "3 and 2 are 5." With the forefinger slide down the Heaven counter on the same upright as the four Earth counters and replace 2 of the Earth counters.

The sum 7 is indicated on the board.

Example 18. 3 and 3 are 6.

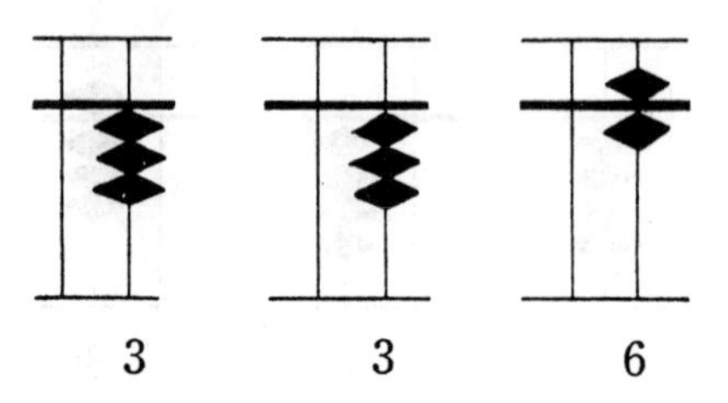

(1) With the thumb raise three Earth counters.

(2) Think "3 and 2 are 5." Slide down the Heaven counter on the same upright as the three Earth counters and at the same time with the forefinger replace 2 of the Earth counters.

The sum 6 is indicated on the board.

Example 19. 2 and 3 are 5.

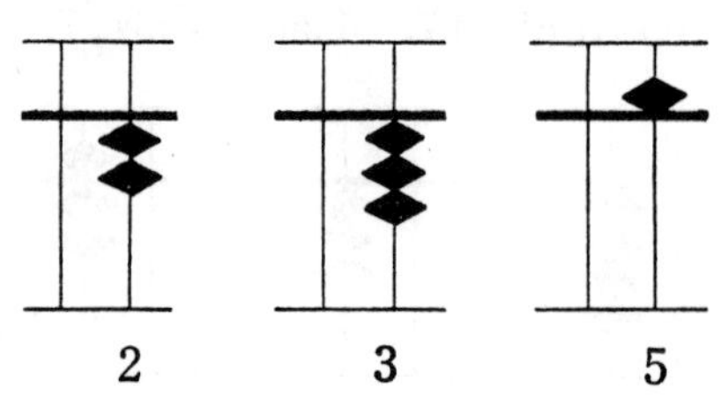

(1) With the thumb raise two of the Earth counters.

(2) Think, "2 and 3 are 5" and with the forefinger slide down the 5 counter on the same upright as the two Earth counters. At the same time return to place the two Earth counters.

The sum 5 is indicated on the board.

Example 20. 4 and 2 are 6.

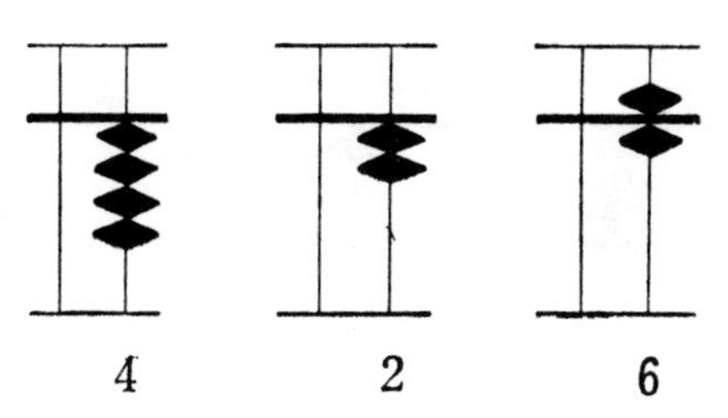

(1) With the thumb raise four of the one counters.

(2) Think, "2 and 3 are 5." With the forefinger take down the 5 counter on the same upright with the four one counters and at the same time replace three of the one counters.

The sum 6 is indicated on the board.

Example 21. 3 and 2 are 5.

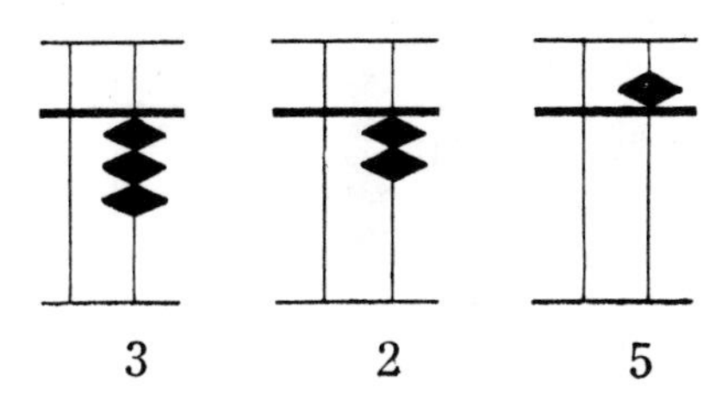

(1) With the thumb raise three of the one counters into place.

(2) Think, "2 and 3 are 5." With the forefinger slide down the 5 counter on the same upright and at the same time return to their original position the three one counters.

The sum 5 is indicated on the board.

Example 22. 4 and 1 are 5.

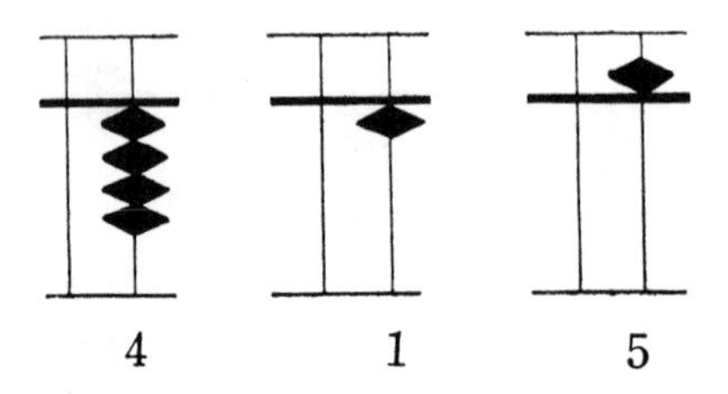

(1) With the thumb raise four of the one counters.

(2) Think, "1 and 4 are 5." With the forefinger take down the 5 counter on the same upright as the four counters and at the same time return the four counters to place.

The sum 5 is indicated on the board.

Example 23. 1 and 9 are 10.

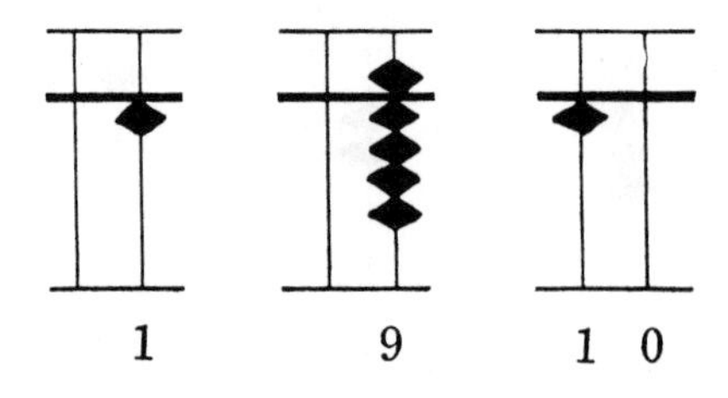

(1) With the thumb raise one Earth counter.

(2) Think, "A. T. 1 and 9 are 10." With the forefinger replace the one counter and at the same time with the thumb raise one of the Earth counters on the next upright to the left.

The sum 10 is indicated on the board.

Example 24. 2 and 8 are 10.

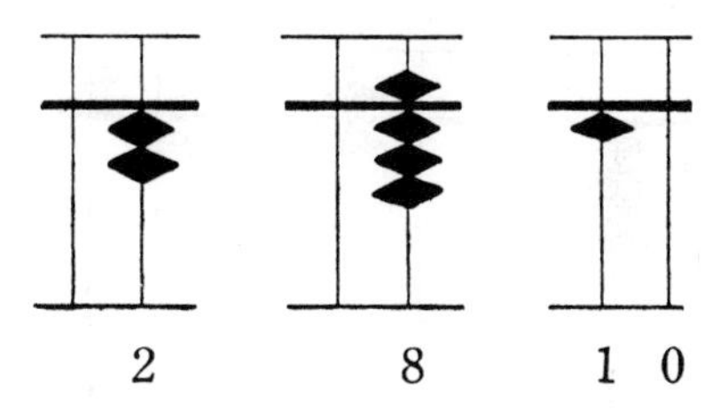

(1) With the thumb raise two of the Earth counters.

(2) Think, "A. T. 2 and 8 are 10." With the forefinger replace the two Earth counters and at the same time raise one of the Earth counters on the next upright to the left, with the thumb.

The sum 10 is indicated on the board.

Example 25. 3 and 7 are 10.

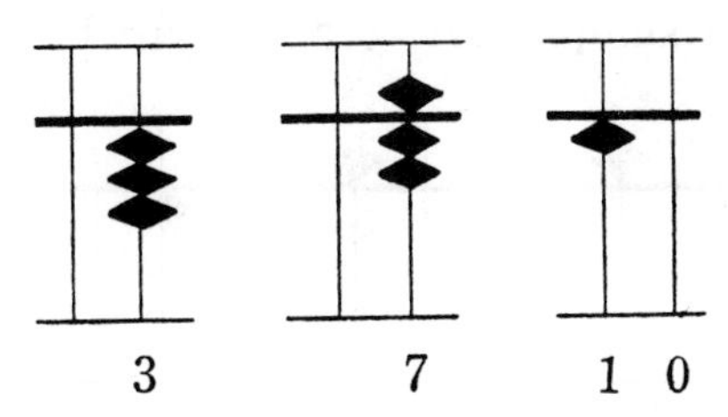

(1) With the thumb raise three of the Earth counters.

(2) Think, "A. T. 3 and 7 are 10." With the forefinger return the three Earth counters and at the same time raise one of the Earth counters on the next upright to the left, with the thumb.

The sum 10 is indicated on the board.

Example 26. 4 and 6 are 10.

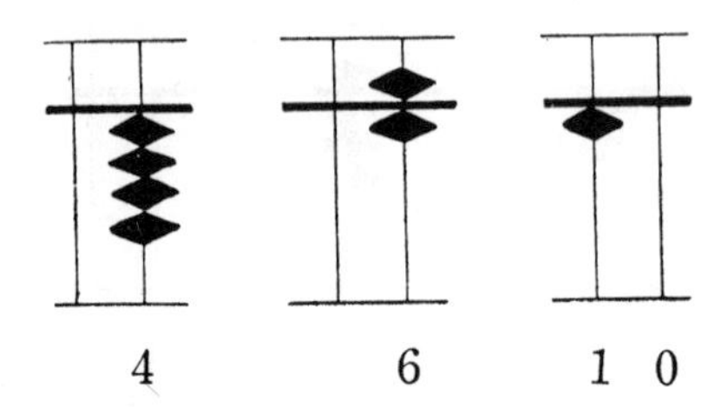

(1) With the thumb raise four Earth counters.

(2) Think, " A. T. 4 and 6 are 10." At one time replace the four Earth counters with the forefinger and raise one Earth counter on the next upright to the left, with the thumb.

The sum 10 is indicated.

Example 27. 5 and 5 are 10.

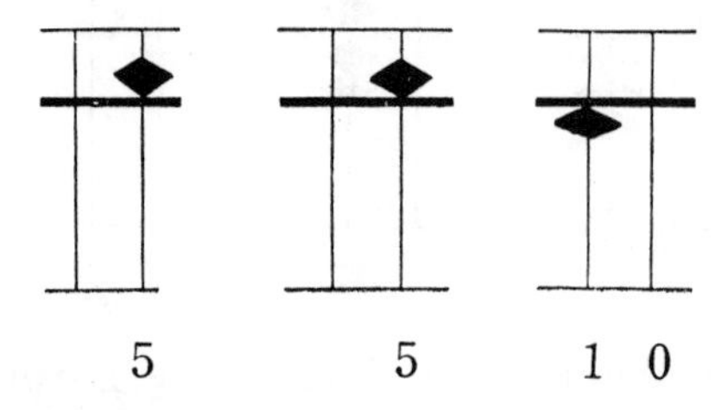

(1) With the forefinger slide down a Heaven counter.

(2) Think, " A. T. 5 and 5 are 10." With the forefinger replace the Heaven counter and at the same time raising one Earth counter on the next upright to the left, with the thumb.

The sum 10 is indicated.

Example 28. 5 and 6 are 11.

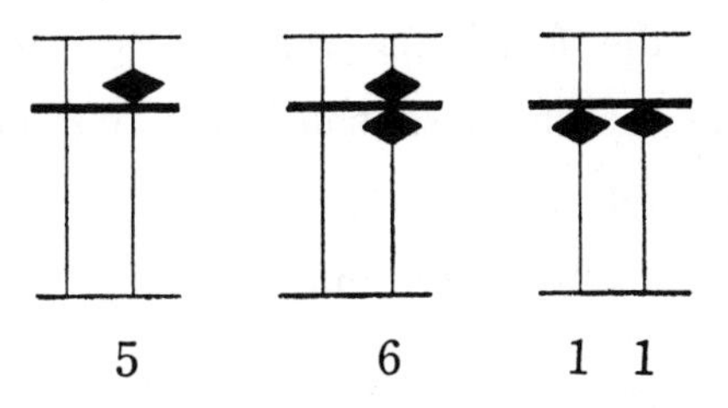

(1) With the forefinger slide down a Heaven counter.

(2) Think, "A.T. 5 and 5 are 10." Raise one Earth counter on the same upright as the Heaven counter which is in position. Do this with the thumb. With one movement return the Heaven counter with the forefinger and with the thumb raise one Earth counter on the next upright to the left.

Eleven is indicated. It is the sum.

Example 29. 5 and 7 are 12.

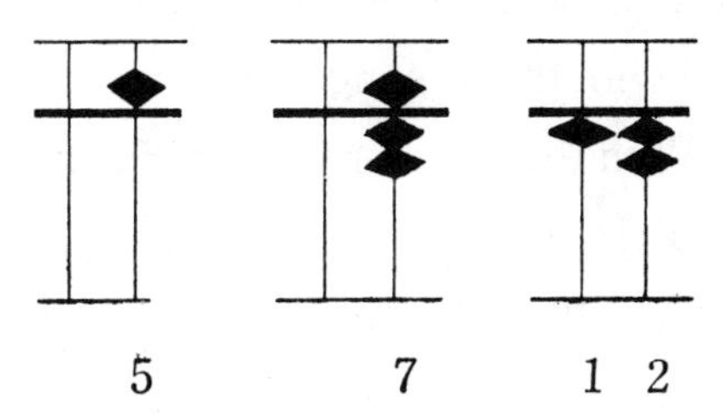

(1) With the forefinger slide down one of the Heaven counters.

(2) Think, "A.T. 5 and 5 are 10." Raise two Earth

counters on the same upright as the Heaven one with the thumb. With the forefinger replace the Heaven counter and with the thumb raise one Earth counter on the next upright to the left.

12 is indicated. It is the sum.

Example 30. 5 and 8 are 13.

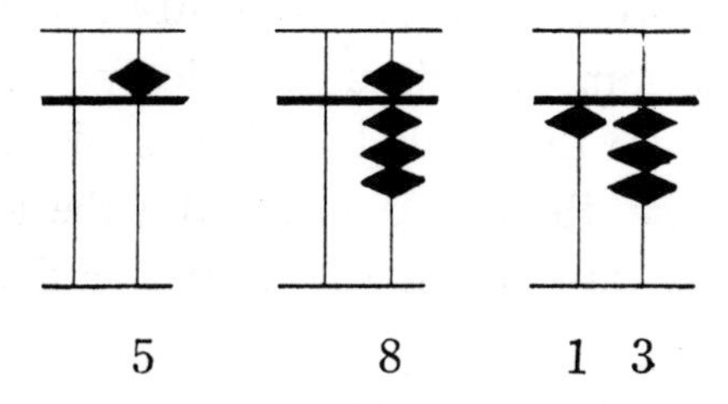

(1) With the forefinger slide down a Heaven counter.

(2) Think, "A. T. 5 and 5 are 10." Raise three Earth counters on the same upright as the Heaven counter with the thumb and with the forefinger return the Heaven counter to its place. With the thumb raise one Earth counter on the next upright to the left.

13 is indicated. It is the sum.

Example 31. 6 and 6 are 12.

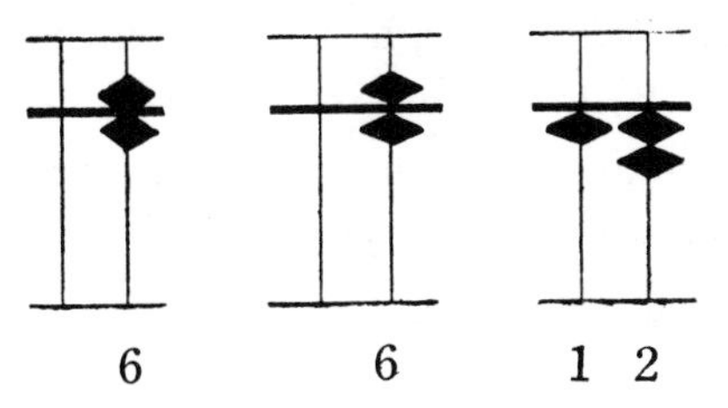

(1) With one movement hold between the forefinger and the thumb one Heaven counter and one Earth counter on the same upright. This will indicate 6.

(2) Think, "A. T. 5 and 5 are 10." Raise one Earth counter on the same upright with the other two with the thumb. With the forefinger replace the Heaven counter and at the same time raise one Earth counter on the next upright to the left, with the thumb.

12 is indicated on the board. It is the sum.

Example 32. 7 and 6 are 13.

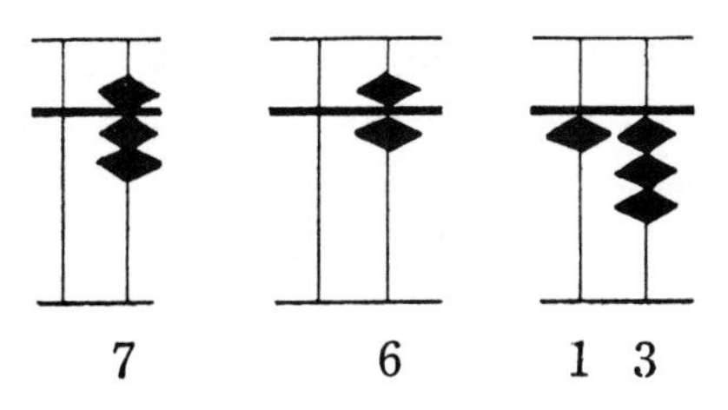

(1) Hold between your forefinger and your thumb one Heaven counter and two Earth counters.

(2) Think, "A. T. 5 and 5 are 10." Raise one Earth counter on the same upright as those in position, with the thumb. With the forefinger return to its original place the Heaven counter and with the thumb raise one Earth counter on the next upright to the left.

13 is indicated. It is the sum.

Example 33. 8 and 6 are 14.

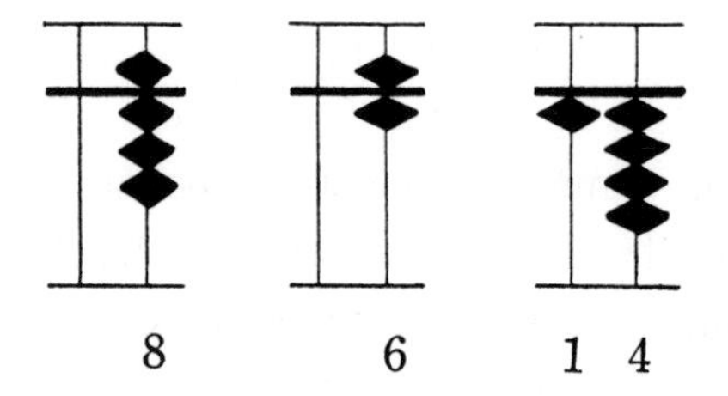

(1) Hold between your thumb and forefinger one Heaven counter and three Earth counters on the same upright.

(2) Think, "A. T. 5 and 5 are 10." Raise one Earth counter up with the other two and with the forefinger replace the Heaven counter. With the thumb raise one Earth counter on the next upright to the left.

14 is indicated. It is the sum.

Example 34. 9 and 9 are 18.

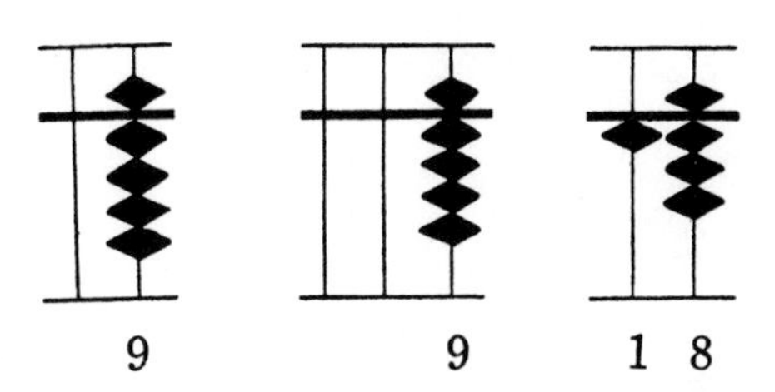

(1) Press between the forefinger and the thumb one Heaven counter and four Earth counters on the same upright.

(2) Think, "A. T. 9 and 1 are 10." Replace one Earth counter which was in position, with the forefinger and at the same time raise one Earth counter on the next upright to the left, with the thumb.

18 is indicated. It is the sum.

Exercise.

Add each column and write the sum.

Row A.	2	3	4	5	7
	3	4	5	6	7
	___	___	___	___	___
Row B,	8	9	3	5	7
	9	1	5	4	3
	___	___	___	___	___

Row C.	2	3	6	7	8
	4	6	6	7	8
Row D.	7	3	9	3	2
	8	3	9	9	7
	9	6	5	5	6
	2	8	5	8	8
	4	3	2	1	2
	4	8	7	8	5

When you use the abacus you must constantly think of the Addition Table even though your numbers are small so that you will form a habit which will be helpful when you begin to handle large numbers.

Section 2.

Part 2.

The Determination of the Position on a Board.

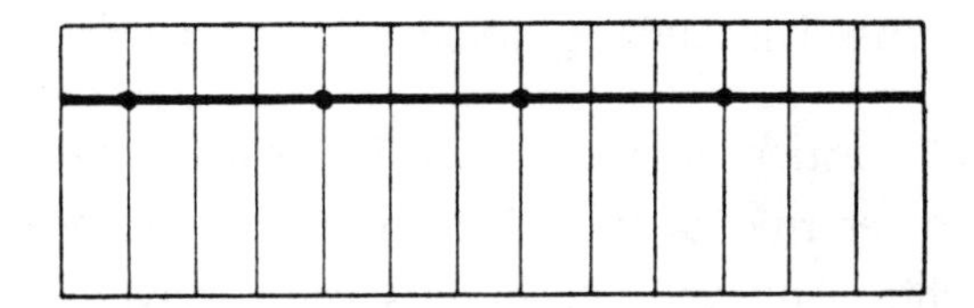

For the sake of convenience the abacus is divided into sections by black spots on the partition between Heaven and Earth.

From one of these black spots one step towards the left is the position for 10's and the next place is for 100's and the next is for 1000's.

For example:

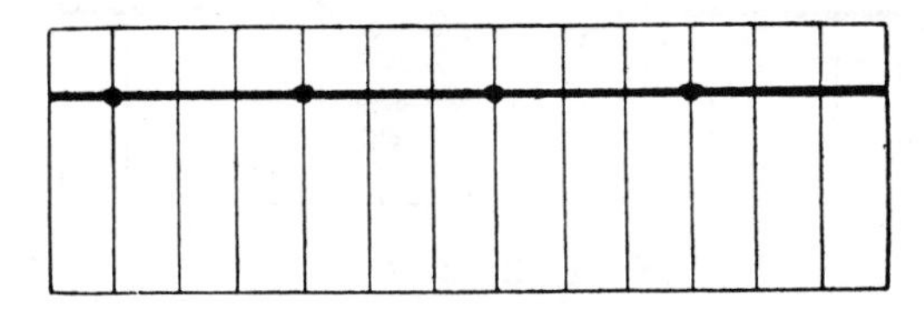

F E D C B A X Y Z

If we decide to use the upright A as the unit position, B is the position for 10's.

C is for 100's
D is for 1000's
E is for 10,000's
F is for 100,000's and so on
X is for one tenth's position
Y is for one hundred's position
Z is for one thousand's position and so on.

The unit position A was chosen in the above example but there is no set rule as to which upright should become the unit upright. Any one which suits your convenience may be chosen, but it should be one with a black spot.

The black spot is very convenient in counting money in the United States. It serves as the decimal point to divide the dollars from the cents.

For example:

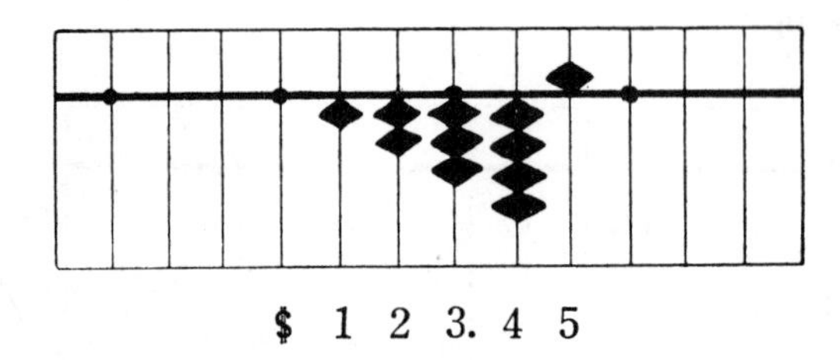

$ 1 2 3. 4 5

The black spot shows that the 123 is dollars and that the 45 is cents.

Section 2.

Part 3.

Illustration of large numbers on a Board.

Now that you understand the determination of a position, chose one upright which is marked by a black dot to serve as the unit position. Then just as in writing a number begin at the left and place the counters into position as needed. Illustrations will make this point clear.

Example 1.

Designate 112.

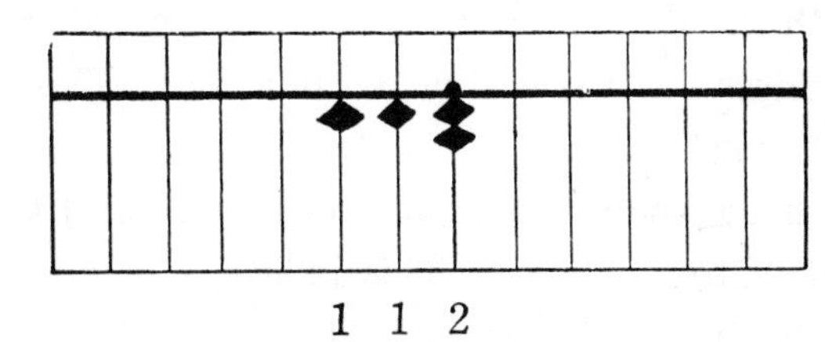

(1) With the thumb raise one Earth counter on the second upright from the one with the black dot. This will indicate one hundred.

(2) With the thumb raise one Earth counter on the first upright to the left. This indicates one ten.

(3) With the thumb raise two Earth counters on the unit upright.

This indicates two ones and the whole indicates 112.

Example 2.

Indicate 2,345.

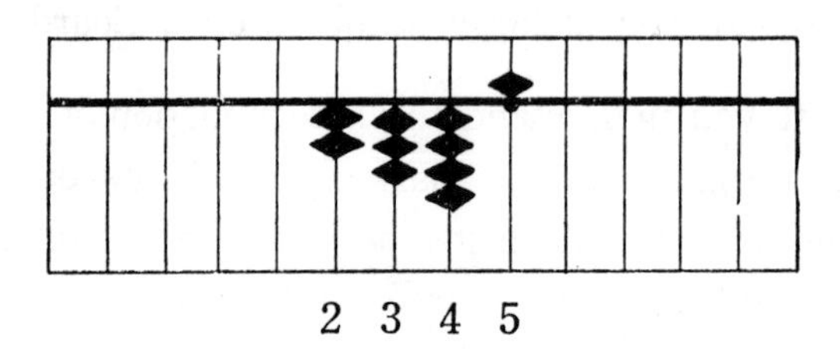

2 3 4 5

(1) With the thumb raise two Earth counters on the third upright to the left of the black spot. This indicates two thousands.

(2) With the thumb raise three Earth counters on the second upright to the left of the unit one. This indicates three one hundreds.

(3) With the thumb raise four counters on the upright on the left of the unit one. This indicates four tens.

(4) With the forefinger slide down the Heaven counter on the unit upright. This indicates five ones.

Altogether 2,345 is indicated.

Section 2.

Part 4.

Adding With Large Numbers.

Adding with numbers of two places or more is the same as adding with numbers less than 10. It is a series of repetition of one place numbers, the only difference being that whenever an upright indicates its full capacity of ten then you must replace all of these counters and raise one Earth counter on the next upright to the left.

Example 1. 56 and 56 are 112.

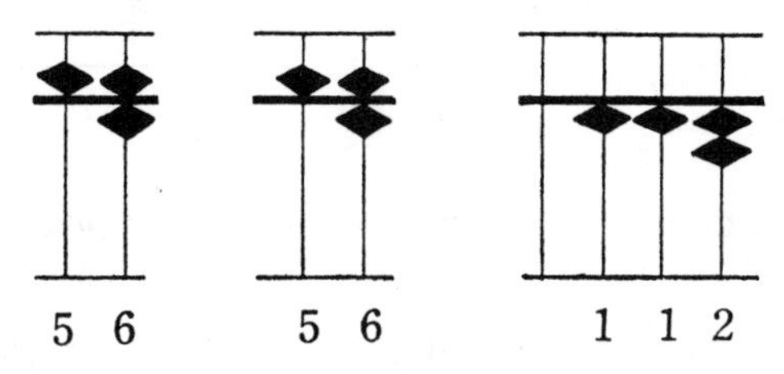

(1) With the forefinger slide down one Heaven counter cn the ten upright in relation to the unit counter.

(2) Press between your fingers the Heaven counter and one Earth counter on the unit upright.

56 is indicated on the board.

(3) Think, "A. T. 5 and 5 are 10." Replace the Heaven counter on the ten upright with the forefinger and at the same time raise one of the Earth counters on the next upright to the left with your thumb. 106 is indicated.

(4) Think, "A. T. 5 and 5 are 10." Raise an Earth counter on the unit upright with the thumb and with the forefinger replace the 5 counter on the same upright. Then raise one Earth counter on the next upright to the left with the thumb.

112 is indicated. It is the sum.

Example 2. 55 and 11 are 66.

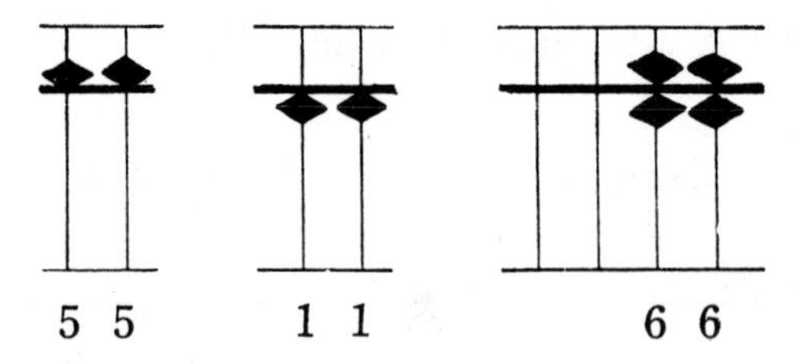

(1) Take down a 5 counter on the ten upright, with the forefinger.

(2) Take down a 5 counter on the unit upright, with the forefinger.

55 is indicated.

(3) Raise one Earth counter on the 10 upright with the thumb.

(4) Raise one Earth counter on the unit upright with the thumb.

66 is indicated on the board. It is the sum.

Example 3. 26 and 42 are 68.

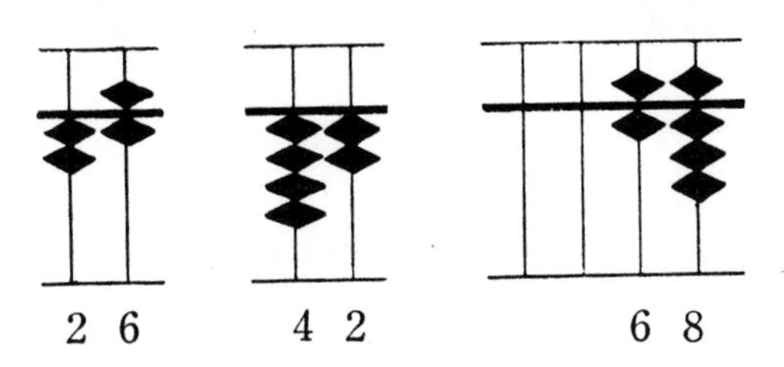

(1) With the thumb raise two Earth counters on the 10 upright.

(2) Press between the forefinger and the thumb the Heaven counter and one Earth counter on the unit upright.

26 is indicated on the board.

(3) Think, "4 and 1 are 5." Slide down the 5 counter on the 10 upright with the forefinger, and at the same time return one Earth counter on the same upright.

(4) Raise two Earth counters on the unit upright with the thumb.

68 is indicated on the board. It is the sum.

Example 4. 33 and 55 are 88.

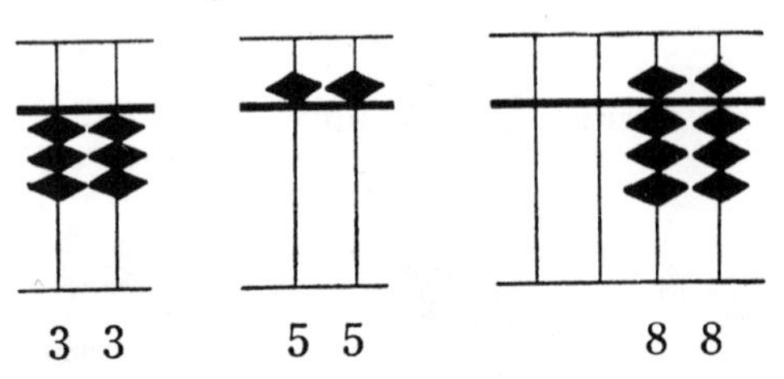

(1) Raise three Earth counters on the 10 position, with the thumb.

(2) Raise three Earth counters on the unit upright, with the thumb.

33 is indicated.

(3) Slide down the 5 counter on the 10 upright with the forefinger.

(4) Slide down the 5 counter on the unit upright with the forefinger.

88 is indicated. It is the sum.

Example 5. 66 and 55 are 121.

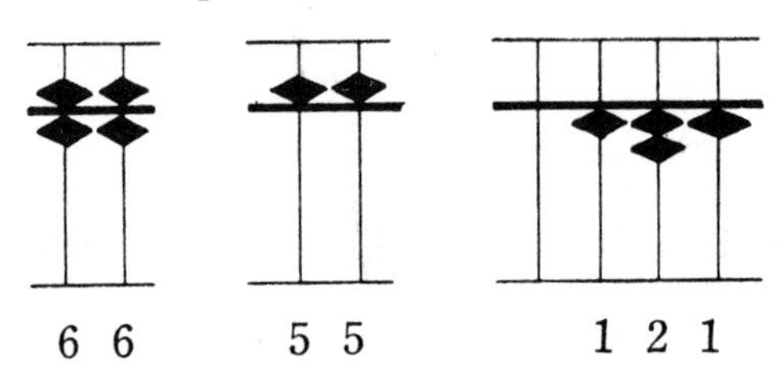

(1) Press between the forefinger and the thumb the 5 counter and one Earth counter on the 10 upright.

(2) Hold between the forefinger and the thumb the 5 counter and one Earth counter on the unit upright.

66 is indicated on the board.

(3) Think, "A. T. 5 and 5 are 10." With the forefinger return to position one Heaven counter on the 10 upright, and at the same time raise one Earth counter on the next upright to the left, in this case the 100 position.

116 is indicated on the board.

(4) Think, "A. T. 5 and 5 are 10." With the forefinger replace the Heaven counter on the unit position, and at the same time raise one Earth counter on the next upright to the left, with the thumb.

121 is indicated on the board. It is the sum.

Example 6. 34 and 89 are 123.

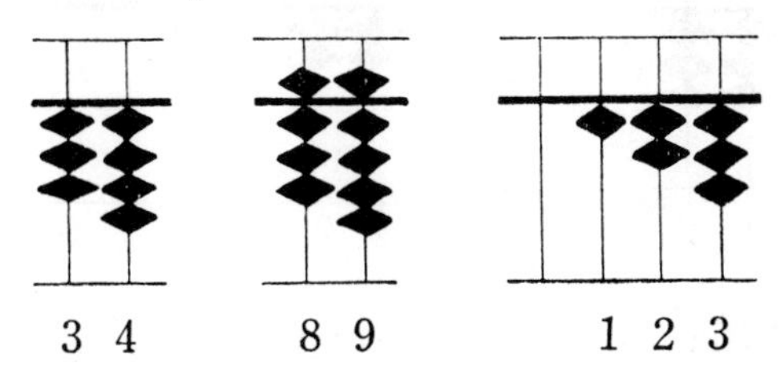

(1) Raise three Earth counters on the 10 upright with the thumb.

(2) Raise four Earth counters on the unit upright with the thumb.

34 is indicated on the board.

(3) Think, "A. T. 8 and 2 are 10." With the forefinger return two Earth counters on the 10 upright, and at the same time, with the thumb, raise one Earth counter on the next upright to the left.

114 is indicated on the board.

(4) Think, "A. T. 9 and 1 are 10." With the forefinger return one Earth counter on the unit upright, and at the same time with the thumb, raise one Earth counter on the next upright to the left.

123 is indicated on the board. It is the sum.

Example 7. 333 and 222 are 555.

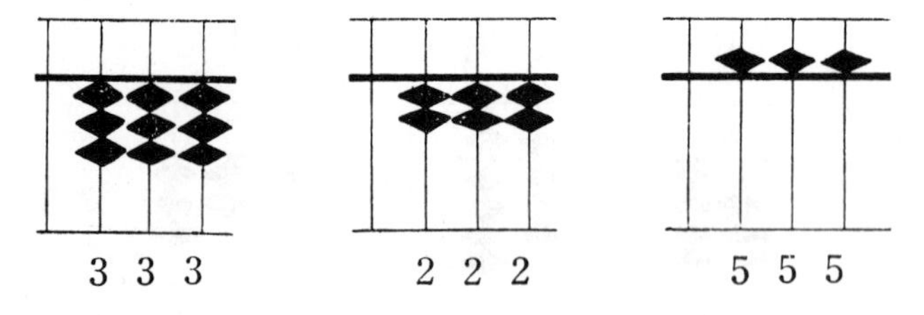

(1) With the thumb raise three Earth counters on the 100 upright.

(2) With the thumb raise three Earth counters on the 10 upright.

(3) With the thumb raise three Earth counters on the unit upright.

333 is indicated on the board.

(4) Think, "2 and 3 are 5." Slide down the Heaven counter on the 100 upright with the forefinger, and at the same time return to place three Earth counters on the same upright.

(5) Think, "2 and 3 are 5." Slide down the Heaven counter on the 10 upright with the forefinger, and at the same time return three Earth counters on the same upright.

(6) Think, "2 and 3 are 5." Slide down the Heaven counter on the unit upright with the forefinger and at the same time return three Earth counters on the same upright.

The sum 555 is indicated on the board.

Example 8. 234 and 999 are 1233.

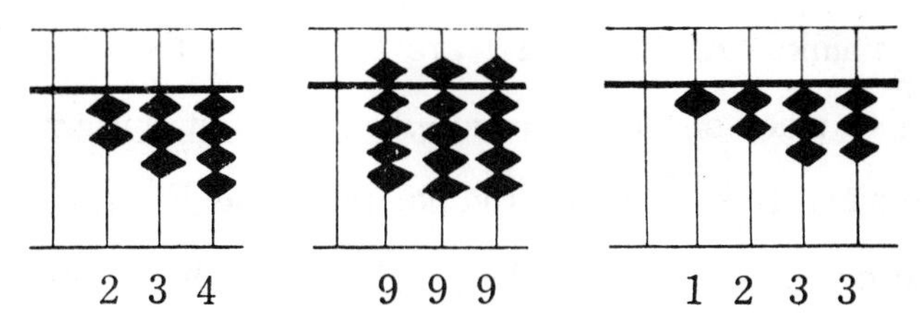

(1) With the thumb raise two of the Earth counters on the 100 upright.

(2) With the thumb raise three Earth counters on the 10 upright.

(3) With the thumb raise four Earth counters on the unit upright.

234 is now indicated on the board.

(4) Think, "A. T. 9 and 1 are 10." With the forefinger return one Earth counter on the 100 upright, and at the same time with the thumb raise one Earth counter on the next upright to the left, in this case the 1000 upright.

(5) Think, "A. T. 9 and 1 are 10." With the forefinger return to place one Earth counter on the 10 upright, and at the same time raise one Earth counter on the next upright to the left, with the thumb, in this case the 100 upright.

(6) Think, "A. T. 9 and 1 are 10." With the forefinger return to place one Earth counter on the unit upright, and at the same time with the thumb, raise one Earth counter on the next upright to the left, in this case the 10 upright.

1233 is indicated on the board. It is the sum.

As is shown in the above examples if we remember how to add small numbers on the abacus, and remember the Addition Table we can add numbers of any size.

It is usually more important to do things well than to do them quickly but speed is useful and it will prove worth while to learn to handle the abacus quickly and with accuracy.

It is also useful to know how quickly you can do things and still do them well so time yourself when taking tests, not only in addition but in multiplication as well.

Exercise.

Add and write the sum.

Row A.	22 88	66 11	78 21	56 43	45 66
Row B.	11 99	33 99	33 88	567 128	3984 8697
Row C.	22 66	77 11	11 22	11 44	77 67
Row D.	66 66	99 22	64 64	44 77	133 345

Adding large numbers is just as simple as the work which you have just been doing.

Row E.

1465	2564	6315	79	91235
8871	1121	572	452	3492
3296	980	1805	4597	515

Row F.

$50.00	$41.93	$15.14	$53.98
6.71	7.76	19.62	4.72
13.25	11.32	31.58	12.63
31.82	90.11	44.33	25.56

Row G.

28	717	42	57	210
191	238	96	41	551
612	112	123	220	75
233	98	19	309	193
425	216	432	167	61
77	9	111	413	250
339	341	302	700	170

As I have said before, it is not enough to understand the method only, you must acquire speed. You must strive towards what we call "a quick addition."

Section 3.

Addition from Dictation.

The success of adding on the abacus while someone dictates is dependent upon the manner in which the numbers are read out.

The following things should be kept in mind by the reader.

(1) Speak clearly and quickly but in a sing-song manner.

(2) Use some connecting word such as "and" or "next" between the numbers.

(3) If great speed is needed the connecting word should be omitted.

(4) Before reading a column of figures give some warning such as, "Ready?"

(5) If you make a mistake in reading a number ask the adder to add or subtract the amount necessary to correct the mistake. Or if you are worried for fear some mistake has been made ask to begin all over again.

(6) At the end use some expression which signifies that the work is finished such as, "That's all," or, "What's the sum?"

Practise with the following exercises until the adder can get all the sums correctly in a minute or less. Then the reader must read a column in less than a minute.

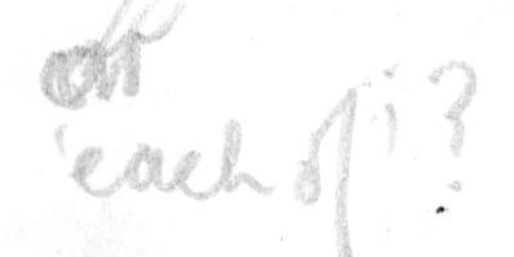

(1)	(2)	(3)	(4)
$ 3.43	$ 7.43	$ 2.76	$ 3.43
7.16	6.89	5.24	7.72
6.88	5.28	1.84	5.94
8.75	6.16	39.63	3.51
1.99	7.24	8.19	2.67
2.37	8.15	3.55	6.76
4.12	9.91	7.31	1.81
8.56	6.78	26.95	9.38
7.19	1.33	4.27	8.53
9.32	2.75	6.98	6.14
4.53	7.64	85.64	4.29
1.79	6.53	9.48	5.32
6.52	8.39	8.37	7.45
8.68	7.51	52.23	7.98
6.53	1.84	4.59	2.61
51.79	92.29	3.12	49.86
79.32	57.64	6.75	51.55
83.87	74.52	81.86	96.43
16.14	56.35	7.43	28.19
65.64	31.57	6.81	74.88
$ 384.58	$ 406.20	$ 373.00	$ 384.45

(5)	(6)	(7)	(8)
$ 7.42	$ 8.93	$ 7.15	$ 3.43
2.99	2.18	2.69	7.72
8.54	9.26	6.98	4.86
5.63	1.35	5.12	8.99
9.88	2.59	1.24	7.72
1.25	7.32	8.71	3.63
6.48	3.77	2.36	4.46
1.31	8.48	9.51	6.37
4.26	1.81	4.95	2.46
3.65	6.64	3.16	1.72
7.43	4.13	1.28	7.31
6.39	9.85	6.43	8.64
5.98	7.61	7.69	3.55
1.52	5.42	5.82	4.89
2.87	2.53	2.56	1.78
73.64	26.79	93.64	68.35
54.71	26.86	51.81	86.11
66.37	75.98	74.48	59.68
18.16	33.15	15.37	92.59
43.12	94.67	43.82	35.23
$ 331.60	$ 339.32	$ 354.77	$ 419.49

(9)	(10)	(11)	(12)
$ 7.29	$ 5.26	$ 7.18	$ 37.85
6.44	27.45	3.55	9.62
91.23	8.33	6.79	5.23
9.61	5.72	25.21	83.14
2.96	3.56	8.66	2.56
8.17	59.22	5.48	4.49
35.35	1.94	7.25	6.77
4.92	2.68	41.38	8.68
3.83	6.13	3.61	51.33
6.51	84.91	9.25	2.14
5.84	7.88	76.49	8.25
2.93	4.37	4.63	5.97
19.59	3.91	8.92	6.64
1.47	6.88	96.55	29.73
8.17	61.19	7.78	1.85
6.55	9.64	8.54	3.44
73.71	7.37	3.95	64.38
4.86	8.41	51.76	8.72
6.68	12.55	9.39	7.61
67.32	4.76	5.27	5.56
$ 373.43	$ 332.16	$ 391.64	$ 353.96

(13)	(14)	(15)	(16)
$ 6.46	$ 6.46	$ 5.26	$ 7.18
5.22	4.35	27.45	3.55
4.86	3.24	8.33	6.79
8.99	2.18	5.72	25.21
7.72	6.51	3.56	8.66
3.63	9.66	59.22	5.48
4.46	4.73	1.94	7.25
6.37	3.32	2.68	41.38
2.46	5.49	6.13	3.61
1.72	9.17	84.91	9.25
7.31	1.64	7.88	76.49
8.64	8.25	4.37	4.63
3.55	6.78	3.91	8.92
4.89	7.98	6.88	96.55
1.78	1.82	61.19	7.78
68.35	88.53	9.64	8.54
86.11	85.95	7.37	3.95
59.68	67.31	8.41	51.76
92.59	72.82	12.55	9.39
35.23	41.69	4.76	5.27
$ 420.02	$ 437.88	$ 332.16	$ 391.64

Section 4.

Addition by Oneself.

When adding by ourselves we are not dependent upon the accuracy of the one who reads out the numbers. We must learn to see the numbers quickly and transfer them accurately to the abacus.

Practice with this page until you can give all the sums correctly in thirty seconds or less.

(1)	(2)	(3)	(4)	(5)
607	793	812	538	236
158	187	238	662	769
235	775	525	843	582
919	317	413	757	414
670	128	677	432	795
721	304	338	765	521
813	777	672	669	975
585	619	325	334	704
792	531	838	188	813
444	835	180	517	952
5944	5266	5018	5705	6761

$ 7.24	$ 3.83	$ 4.61	$ 8.16
8.35	6.50	3.53	2.57
1.63	9.67	1.86	6.25
2.85	8.18	9.27	9.41
6.89	2.57	1.84	3.24
8.32	6.25	6.89	1.13
6.61	5.19	3.13	7.22
8.87	4.81	8.87	2.65
2.53	3.13	5.92	9.43
8.41	6.53	7.35	6.41
$ 61.70	$ 56.66	$ 53.27	$ 56.47

$ 6.32	$ 2.16	$ 2.18	$ 4.61
2.56	7.77	7.59	3.53
1.12	6.65	1.88	1.86
8.15	3.50	6.35	9.27
2.33	4.68	2.23	1.88
9.59	5.27	5.61	6.89
4.68	9.64	4.16	3.13
5.27	5.61	6.89	1.13
7.97	2.81	2.52	5.92
6.56	8.19	6.93	7.35
$ 54.55	$ 56.28	$ 46.34	$ 45.57

Section 5.

Addition of Figures from Tabulated Forms.

If you wish to add figures which are on separate cards or pages, use the right hand for manipulating the abacus and the left hand to turn the pages.

Sometimes it is convenient to place the pages so that all the numbers can be seen and use a card to cover all except the figures which you wish to add, as is shown in the illustration.

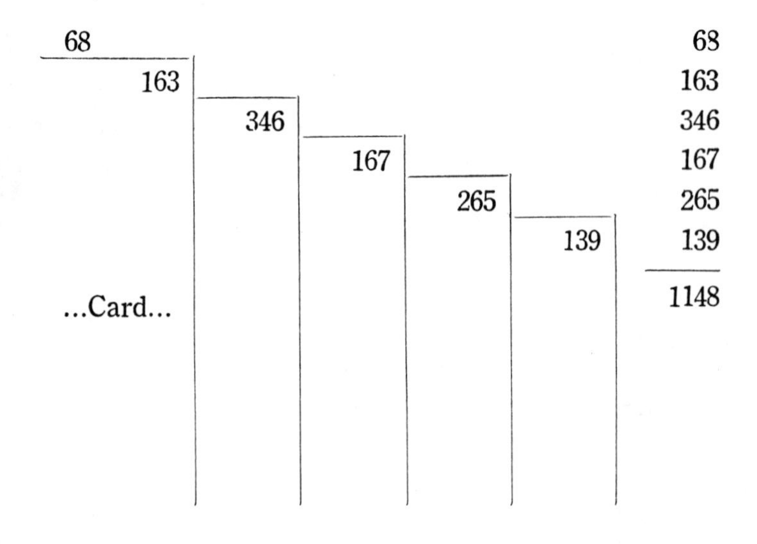

Use the following columns for practice in addition. Find the sums and compare your sums with those given below. Do over any columns which you have done incorrectly.

18	86	24	76	26
56	13	99	59	49
74	26	16	68	66
19	42	43	54	82
84	12	56	22	16
56	11	26	11	40
34	90	27	90	91
24	67	19	34	28
46	37	80	67	56
57	56	47	69	94
34	56	34	28	40
19	17	29	16	69
88	96	50	57	69
72	34	49	96	76
27	23	76	74	53
29	78	54	34	36
33	96	28	43	50
17	43	16	37	96
52	51	76	51	95
65	77	54	73	18
13	18	39	41	34
95	35	24	59	57
1012	1064	966	1159	1241

Add each column.

$ 4.89	$ 19.58	$ 5.95	$ 71.83
6.12	3.32	62.78	8.69
23.58	85.94	1.27	2.71
9.91	4.86	9.31	54.27
85.73	7.79	8.80	1.36
7.04	8.41	33.41	35.04
74.27	41.15	4.66	8.91
6.82	20.36	7.33	9.59
9.65	6.98	96.29	3.17
31.34	4.27	3.16	97.45
3.91	5.43	77.64	6.83
2.56	73.74	8.91	9.32
97.48	8.35	5.55	45.68
8.75	6.69	86.79	2.91
53.93	32.16	5.40	4.49
5.19	9.23	41.14	3.63
8.32	7.02	4.86	19.87
15.81	5.58	58.79	5.10
1.64	64.67	2.72	6.26
6.20	3.98	9.32	51.81
5.98	6.47	93.63	2.18
27.36	33.39	1.40	13.03
6.09	2.12	5.58	8.59
88.47	1.28	68.17	35.95
1.64	50.84	9.26	67.72
14.51	5.66	5.39	3.30
6.75	4.97	23.91	2.41
9.89	69.52	6.53	21.66
65.18	7.95	40.32	6.81
7.33	5.36	2.82	8.25
696.34	607.07	791.09	618.82

Find the sums:

$ 75.53	$ 59.13	$ 16.35	$ 40.52
46.61	14.95	4.97	78.99
1.35	7.38	65.13	9.21
2.20	81.74	9.09	4.82
39.17	2.09	51.42	4.87
2.46	46.65	2.71	5.63
5.42	7.29	5.26	7.18
2.76	6.44	27.45	3.55
1.84	91.23	8.33	6.97
39.63	9.61	5.72	25.21
8.19	2.96	3.56	8.66
3.55	8.17	59.22	5.48
7.31	35.35	1.94	7.25
26.95	4.92	2.69	41.38
4.27	3.83	6.13	3.61
6.98	6.51	84.91	9.25
85.64	5.84	7.88	76.49
9.48	2.93	4.37	4.63
8.39	15.95	3.91	8.92
52.23	1.47	6.88	96.53
4.95	8.12	61.19	7.78
3.12	6.55	9.64	8.54
6.72	73.71	7.37	3.95
81.86	4.86	8.41	51.76
7.34	6.67	12.55	9.39
6.81	67.32	4.89	5.27
37.89	8.64	2.54	8.17
9.62	4.11	3.97	44.66
5.23	23.56	4.89	5.51
83.14	6.93	6.64	7.88
676.64	624.91	500.01	602.06

CHAPTER III.

SUBTRACTION

Subtraction is the process of taking away from something. What is left, or the difference, is called the remainder.

Section 1.

Subtraction Table.

Memorize this table.

$10 - 1 = 9$	One	from	Ten	is	Nine.
$10 - 2 = 8$	Two	from	Ten	is	Eight.
$10 - 3 = 7$	Three	from	Ten	is	Seven.
$10 - 4 = 6$	Four	from	Ten	is	Six.
$10 - 5 = 5$	Five	from	Ten	is	Five.
$10 - 6 = 4$	Six	from	Ten	is	Four.
$10 - 7 = 3$	Seven	from	Ten	is	Three.
$10 - 8 = 2$	Eight	from	Ten	is	Two.
$10 - 9 = 1$	Nine	from	Ten	is	One.

Section 2.

The Practical Use of the Counters
in Subtraction.

You have already learned the method of addition.

Subtraction is just the contrary of addition, but it could be considered the same thing if you think of how much you must add to the subtrahend to make the minuend.

Without knowing it you have already subtracted in doing Example 23 in Section 2 of Addition.

Now I will show you a method of subtraction which is just the same as addition, and if you understand how to handle numbers of one place, it will be easy to calculate the difference between any numbers.

Section 2.

Part 1.

Subtracting with Numbers of One
Place.

Example 1. 1 from 2 is 1.

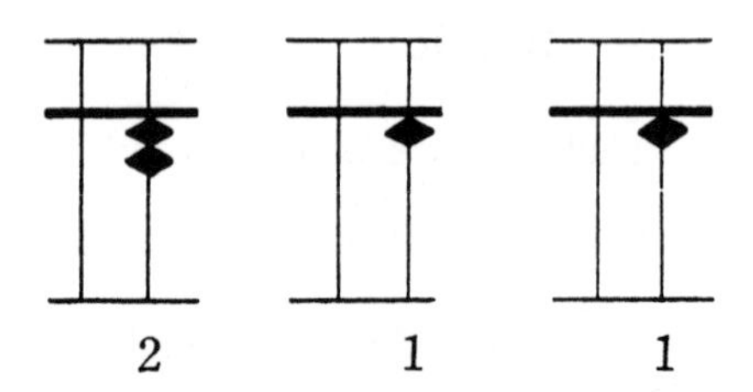

(1) With the thumb raise two Earth counters on the unit position.

(2) Replace one of these with the forefinger.

One is indicated on the board. It is the remainder.

Example 2. 2 from 3 is 1.

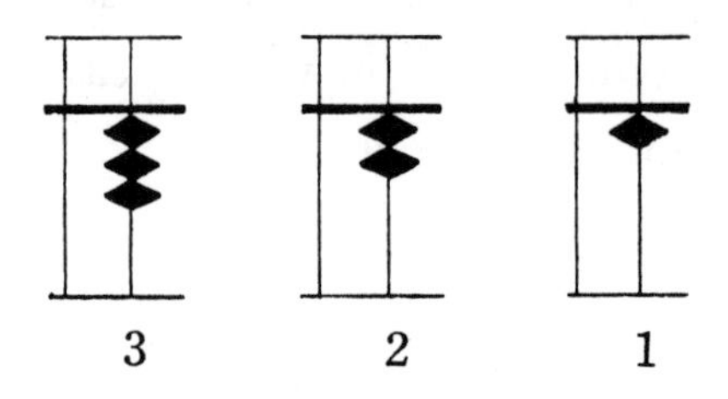

(1) With the thumb raise three Earth counters on the unit upright.

(2) With the forefinger take away two of these.

One is indicated on the board. It is the remainder.

Example 3. 3 from 4 is 1.

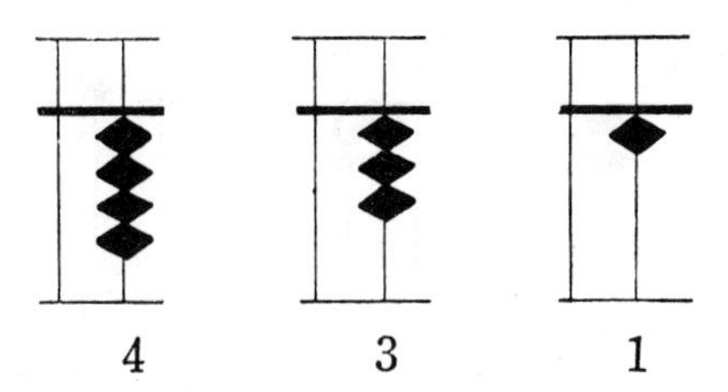

(1) With the thumb raise four Earth counters on the unit upright.

(2) Take away three of these with the forefinger.

One is indicated on the board. It is the remainder.

Example 4. 1 from 6 is 5.

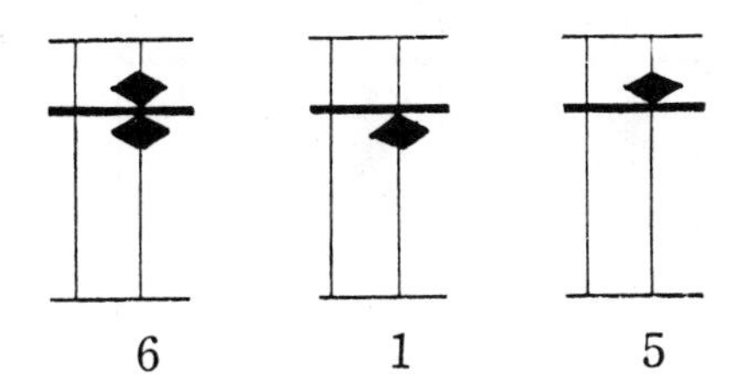

(1) Press between your fingers the Heaven counter and one Earth counter on the unit position.

(2) With the forefinger replace the Earth counter.

Five is indicated on the board. It is the remainder.

Example 5. 2 from 7 is 5.

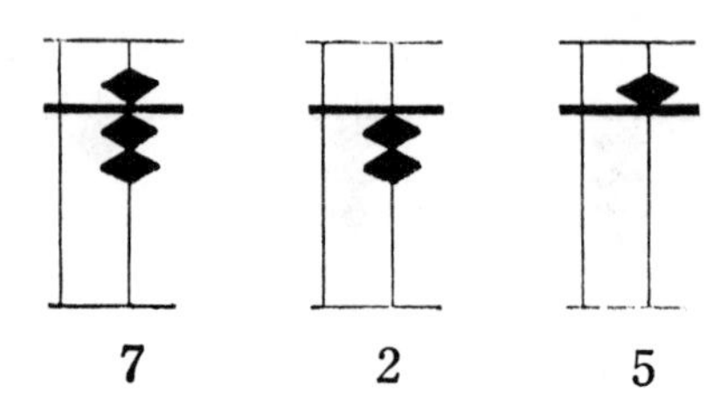

(1) Press between the fingers one Heaven counter and two Earth counters on the unit position.

(2) With the forefinger return the two Earth counters.

5 is indicated on the board. It is the remainder.

Example 6. 3 from 8 is 5.

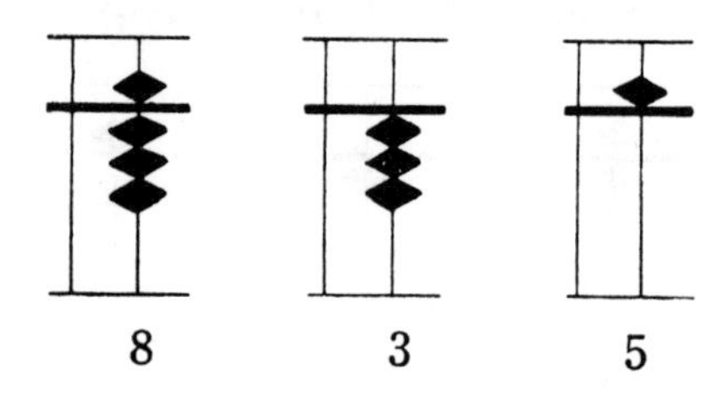

(1) Press between the fingers one Heaven counter and three Earth counters on the unit upright.

(2) Take away the three Earth counters using the forefinger.

5 is indicated on the board. It is the remainder.

Example 7. 4 from 9 is 5.

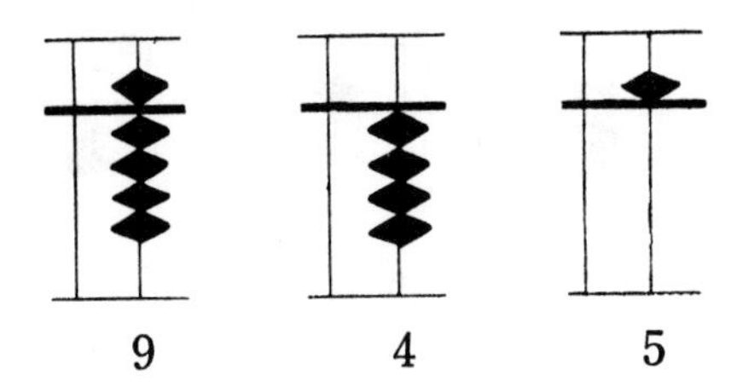

(1) Press between the fingers the Heaven counter and four Earth counters on the unit upright.

(2) With the forefinger take away the four Earth counters.

5 is indicated on the board. It is the remainder.

Example 8. 5 from 6 is 1.

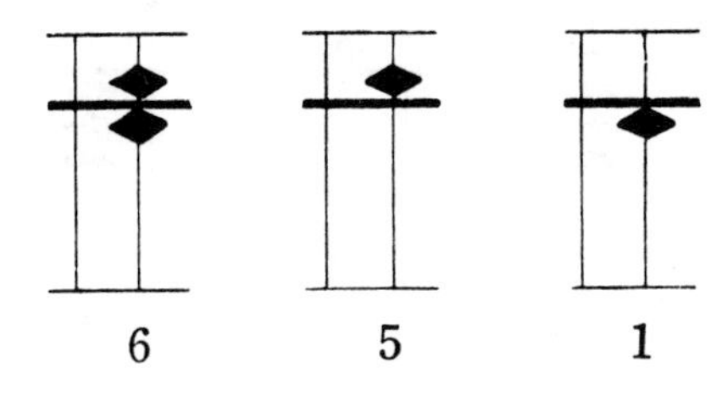

(1) Press between the fingers the Heaven counter and one Earth counter on the unit upright.

(2) Take away the Heaven counter using the forefinger.

1 is indicated on the board. It is the remainder.

Example 9. 5 from 7 is 2.

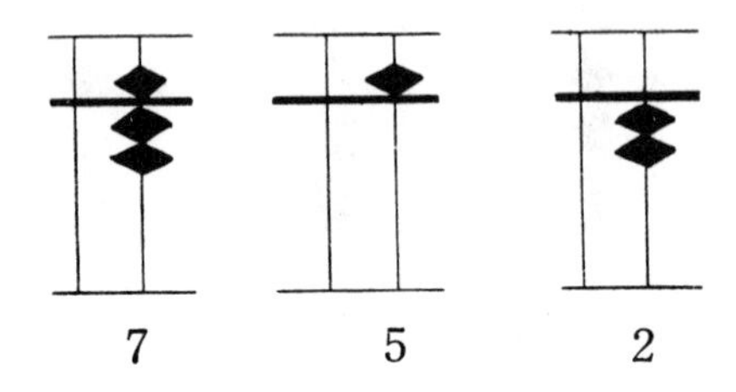

(1) Press between the fingers the Heaven counter and two Earth counters.

(2) With the forefinger replace the Heaven counter.

2 is indicated on the board. It is the remainder.

Example 10. 5 from 8 is 3.

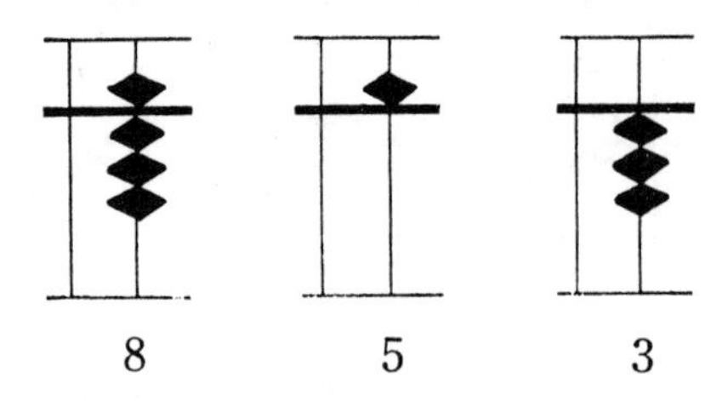

(1) Press between the fingers the Heaven counter and three Earth counters on the unit upright.

(2) With the forefinger replace the Heaven counter.

3 is indicated on the board. It is the remainder.

Example 11. 5 from 9 is 4.

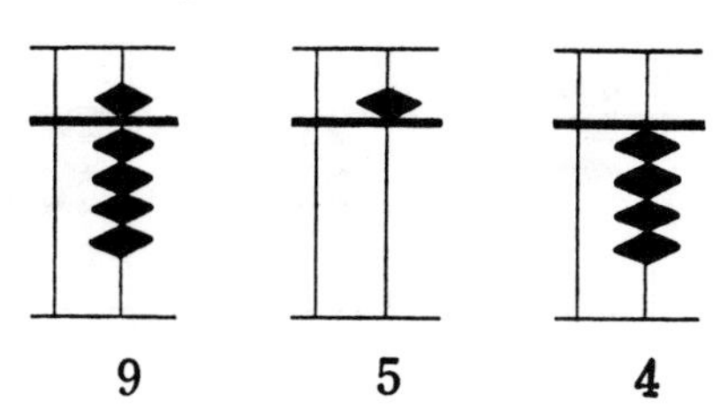

(1) Press between the fingers the Heaven counter and four Earth counters on the unit position.

(2) With the forefinger replace the Heaven counter.

4 is indicated on the board. It is the remainder.

Example 12. 4 from 8 is 4.

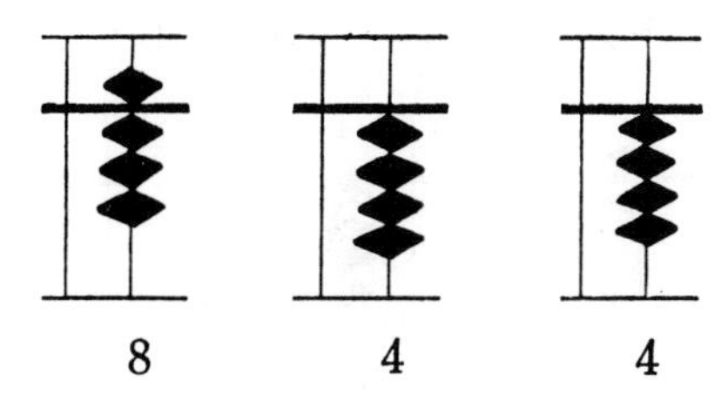

(1) Press between the fingers the Heaven counter and three Earth counters on the unit upright.

(2) Think, "4 from 5 is 1." With the thumb raise one Earth counter on the same upright as the others and at the same time return to place the Heaven counter, using the forefinger.

4 is indicated on the board. It is the remainder.

Example 13. 4 from 7 is 3.

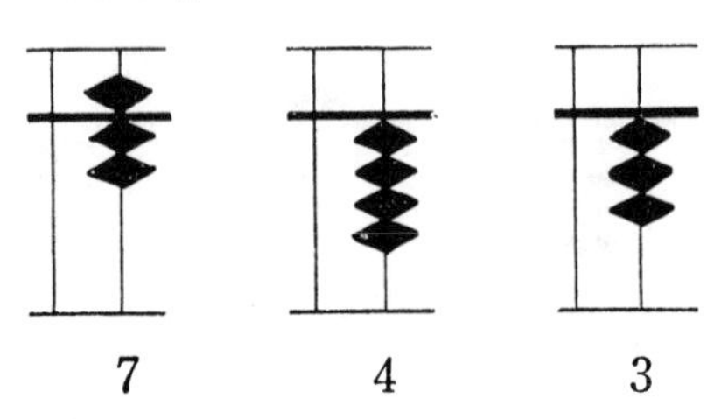

(1) Press between the fingers the Heaven counter and two Earth counters on the unit upright.

(2) Think, "4 from 5 is 1." With the thumb raise one Earth counter on the same upright and at the same time, with the forefinger, return the Heaven counter.

3 is indicated on the board. It is the remainder.

Example 14. 4 from 6 is 2.

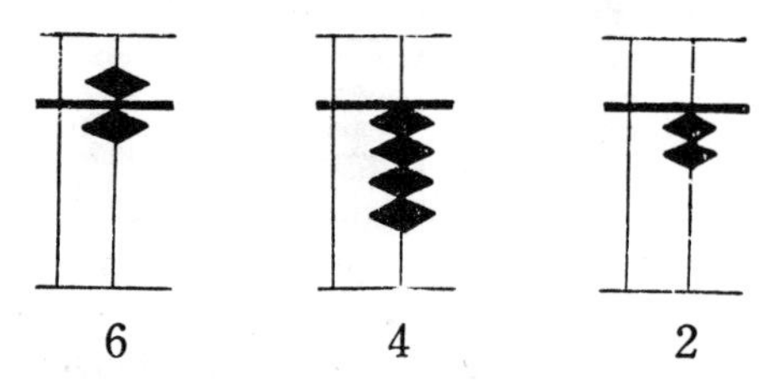

(1) Press between the fingers the Heaven counter and one Earth counter on the unit upright.

(2) Think, "4 from 5 is 1." With the thumb raise one Earth counter on the same upright and at the same time, using the forefinger, return to place the Heaven counter.

2 is indicated on the board. It is the remainder.

Example 15. 3 from 7 is 4.

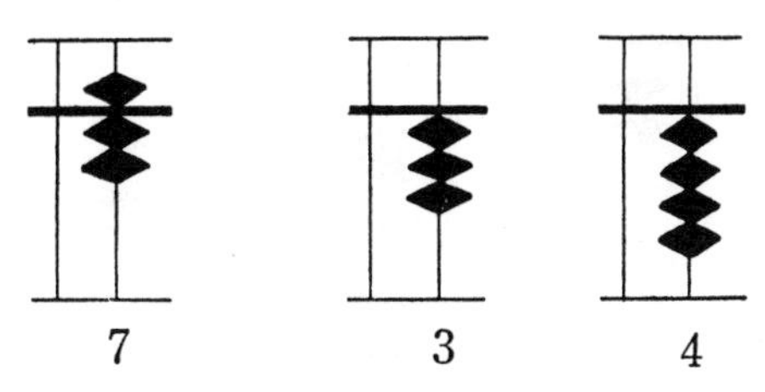

(1) Press between the fingers the Heaven counter and two Earth counters on the unit upright.

(2) Think, "3 from 5 is 2." With the thumb raise two Earth counters on the same upright and, using the forefinger, return the Heaven counter.

4 is indicated on the board. It is the remainder.

Example 16. 3 from 6 is 3.

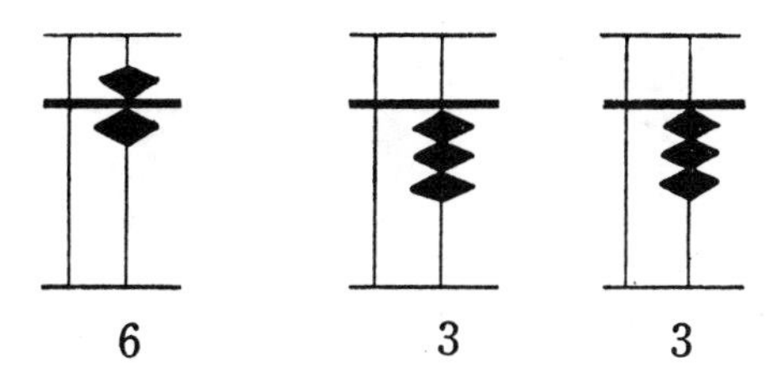

(1) Press between the fingers the Heaven counter and one Earth counter on the unit upright.

(2) Think, "3 from 5 is 2." With the thumb raise two Earth counters on the same upright, and, using the forefinger, replace the Heaven counter.

3 is indicated on the board. It is the remainder.

Example 17. 2 from 6 is 4.

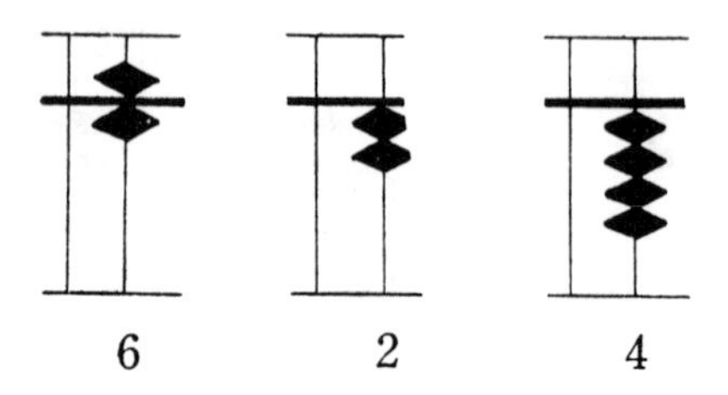

(1) Press between the fingers the Heaven counter and one Earth counter on the unit upright.

(2) Think, "2 from 5 is 3." With the thumb raise three Earth counters on the same upright, and at the same time, using the forefinger, return to place the Heaven counter.

4 is indicated on the board. It is the remainder.

Example 18. 1 from 5 is 4.

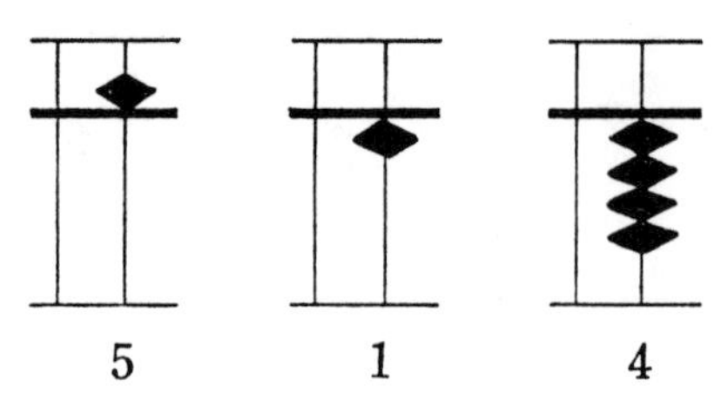

(1) With the forefinger slide down the Heaven counter on the unit upright.

(2) Think, "1 from 5 is 4." With the thumb raise four Earth counters on the same upright and, using the forefinger, return to place the Heaven counter.

4 is indicated on the board. It is the remainder.

Example 19. 2 from 5 is 3.

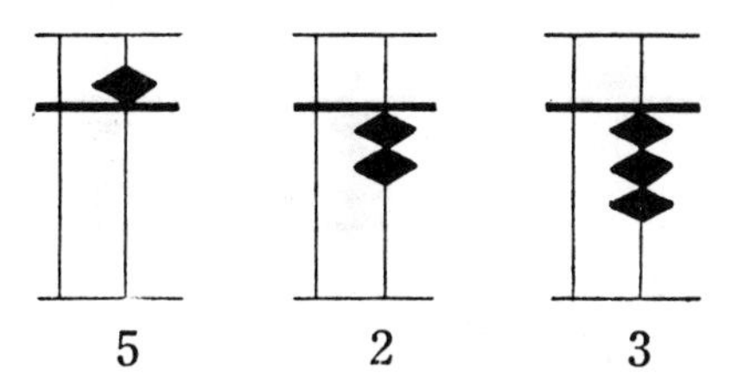

(1) With the forefinger slide down the Heaven counter on the unit upright.

(2) Think, "2 from 5 is 3." With the thumb raise three Earth counters on the same upright and at the same time, using the forefinger, return the Heaven counter to place.

3 is indicated on the board. It is the remainder.

Example 20. 3 from 5 is 2.

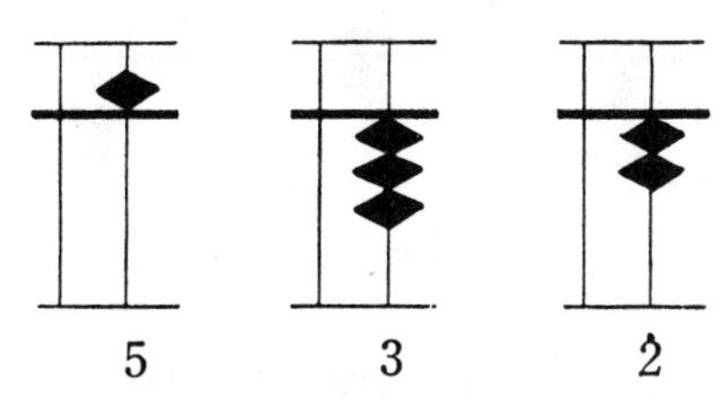

(1) With the forefinger slide down the Heaven counter on the unit upright.

(2) Think, "3 from 5 is 2." With the thumb raise two Earth counters on the same upright and at the same time, with the forefinger, return the Heaven counter to place.

2 is indicated on the board. It is the remainder.

Example 21. 4 from 5 is 1.

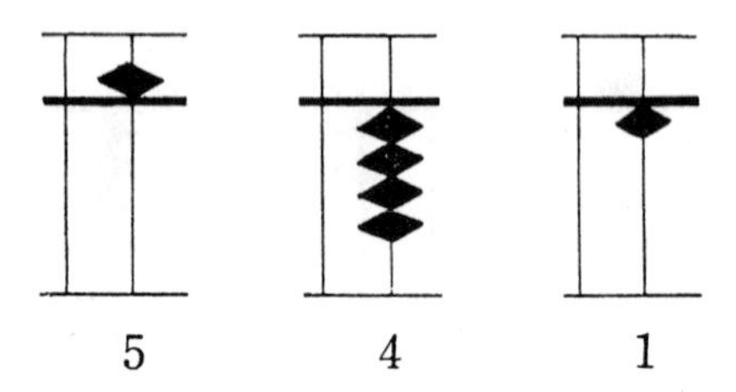

(1) With the forefinger slide down the Heaven counter on the unit upright.

(2) Think, "4 from 5 is 1." With the thumb raise one Earth counter on the same upright, and at the same time, with the forefinger, return the Heaven counter to place.

1 is indicated on the board. It is the remainder.

Example 22. 6 from 9 is 3.

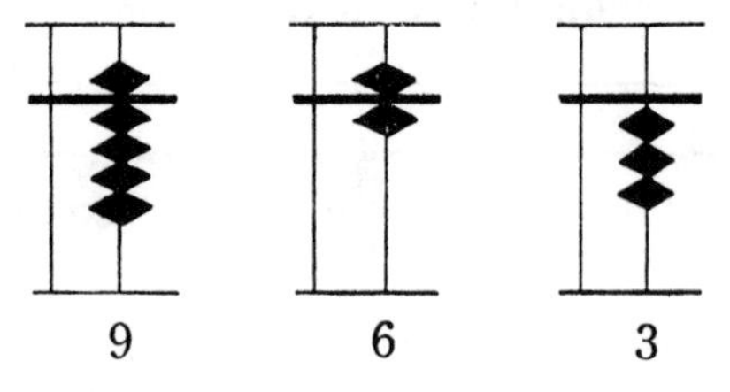

(1) Press between the fingers the Heaven counter and four Earth counters on the unit upright.

(2) At one time return the Heaven counter with the forefinger and one Earth counter with the thumb. The remainder is three.

Notice that in these cases when we return a Heaven counter and an Earth counter at the same time we must use two fingers.

Example 23. 6 from 8 is 2.

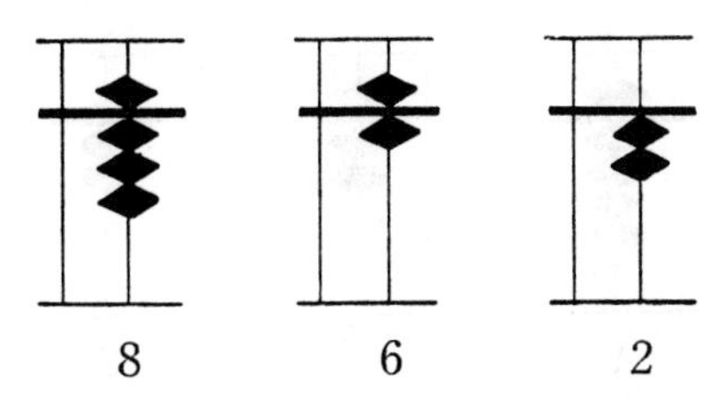

(1) Press between the fingers one Heaven counter and three Earth counters on the unit upright.

(2) Return to place the Heaven counter and one of the Earth counters at the same time, using both forefinger and thumb.

2 is indicated on the board. It is the remainder.

Example 24. 6 from 7 is 1.

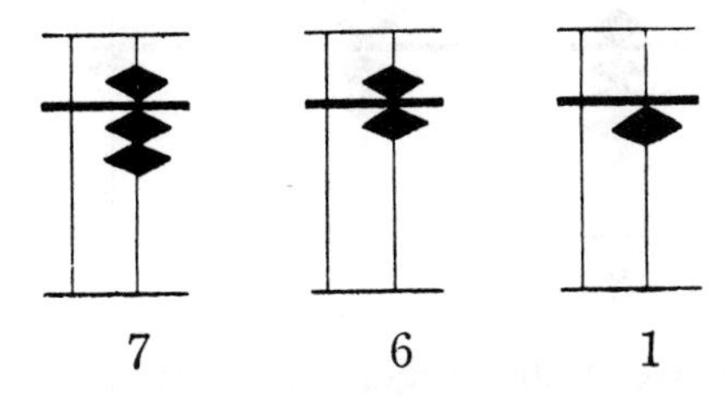

(1) Press between the fingers one Heaven counter and two Earth counters on the unit upright.

(2) Return to place the Heaven counter and one Earth counter at the same time, using both forefinger and thumb.

1 is indicated on the board. It is the remainder.

Example 25. 7 from 9 is 2.

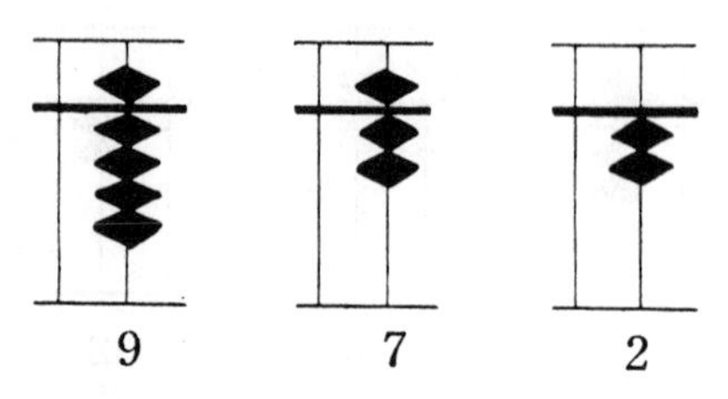

(1) Press between the fingers the Heaven counter and four Earth counters on the unit upright.

(2) Return to place the Heaven counter and two of the Earth counters at the same time, using the forefinger and the thumb.

2 is indicated on the board. It is the remainder.

Example 26. 7 from 8 is 1.

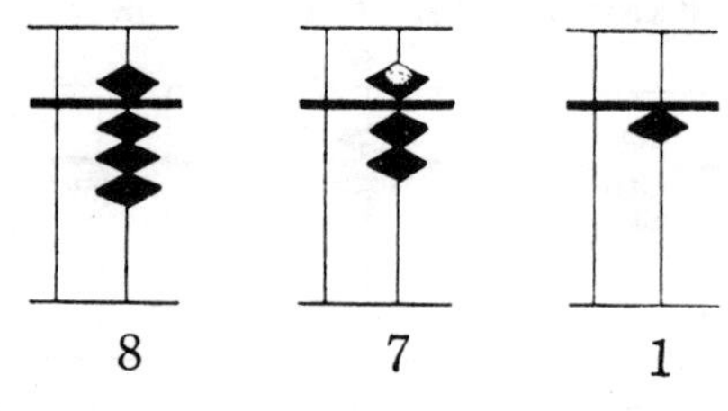

(1) Press between the fingers the Heaven counter and three Earth counters on the unit upright.

(2) Return to place the Heaven counter and two of the Earth counters at the same time, using the forefinger and the thumb.

1 is indicated on the board. It is the remainder.

Example 27. 8 from 9 is 1.

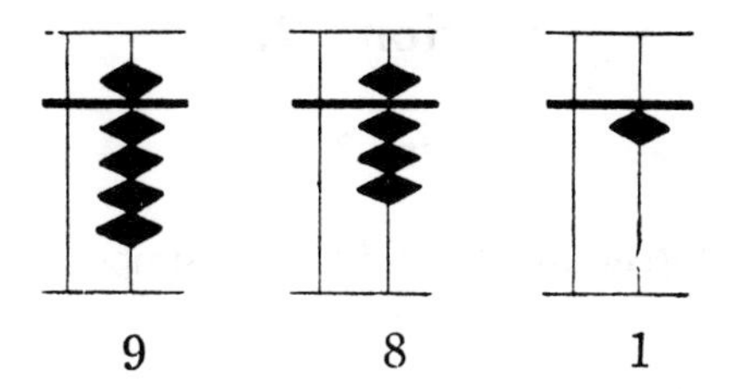

(1) Press between the fingers the Heaven counter and four Earth counters on the unit upright.

(2) Return to place the Heaven counter and three of the Earth counters at one time, using the forefinger and the thumb.

1 is indicated on the board. It is the remainder.

In a general way this completes the method of subtraction.

Although we have dealt with numbers of one place only, using this method as a criterion, we can now subtract larger numbers.

As the method of determining the position on the board, and of handling the counters is practically the same as that of addition, I will abbreviate my explanations.

In these exercises the abbreviation "S. T.", means "Subtraction Tables."

Section 2.

Part 2.

Subtracting with Large Numbers.

Example 1. 1 from 10 is 9.

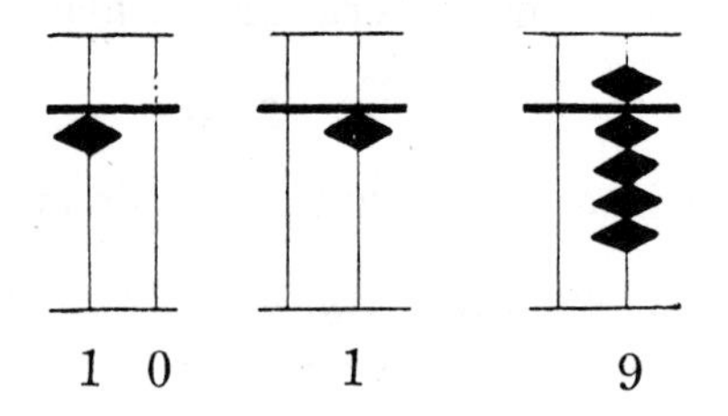

(1) With the thumb raise one Earth counter on the next upright to the left of the unit upright. This indicates ten.

(2) Think, "S. T. 1 from 10 is 9." With the forefinger return the Earth counter on the 10 upright and press between your fingers the Heaven counter and four Earth counters on the unit upright.

9 is indicated on the board. It is the remainder.

Example 2. 2 from 10 is 8.

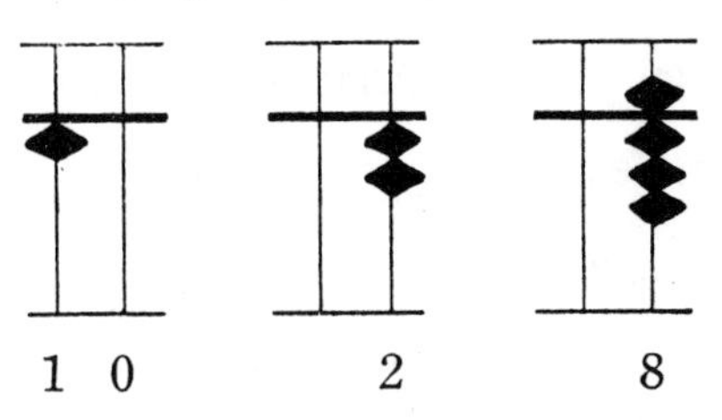

(1) With the thumb raise one Earth counter on the next upright to the left of the unit upright. This indicates 10.

(2) Think, "S. T. 2 from 10 is 8." Replace the Earth counter on the 10 upright, with the forefinger. Then press between your fingers the Heaven counter and three Earth counters of the unit upright.

8 is indicated on the board. It is the remainder.

Example 3. 2 from 11 is 9.

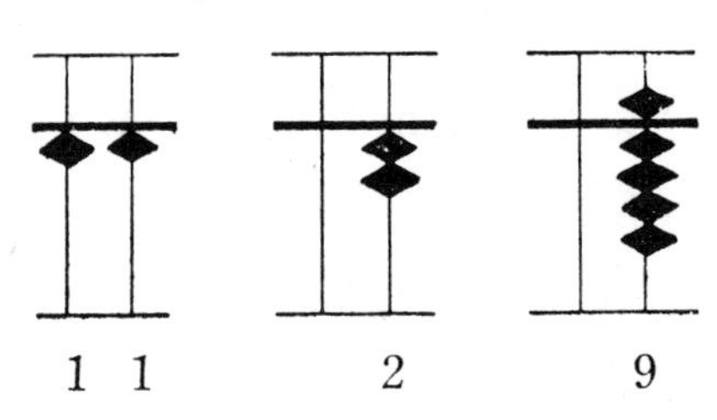

(1) With the thumb raise one counter on the 10 upright.

(2) With the thumb raise one Earth counter on the unit upright. 11 is now indicated on the board.

(3) Think, "S. T. 2 from 10 is 8." Return to place the Earth counter on the 10 upright and then press between the fingers the Heaven counter and three Earth counters on the unit upright.

9 is indicated on the board. It is the remainder.

Example 4. 3 from 10 is 7.

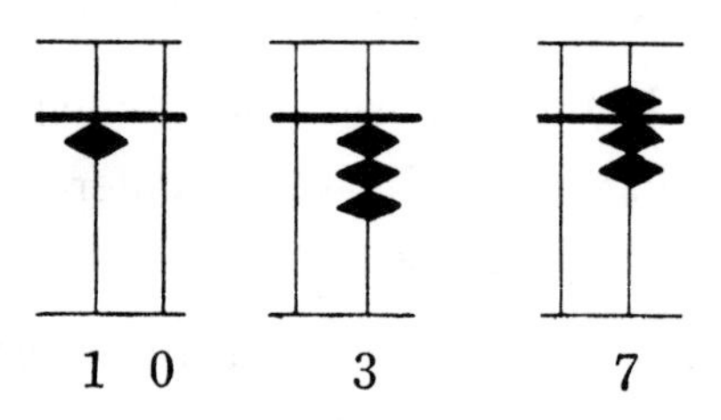

(1) With the thumb raise one Earth counter on the 10 upright.

(2) Think, "S. T. 3 from 10 is 7." With the foretinger replace the Earth counter on the 10 upright and press between the fingers the Heaven counter and three Earth counters on the unit upright.

7 is indicated on the board. It is the remainder.

Example 5. 3 from 11 is 8.

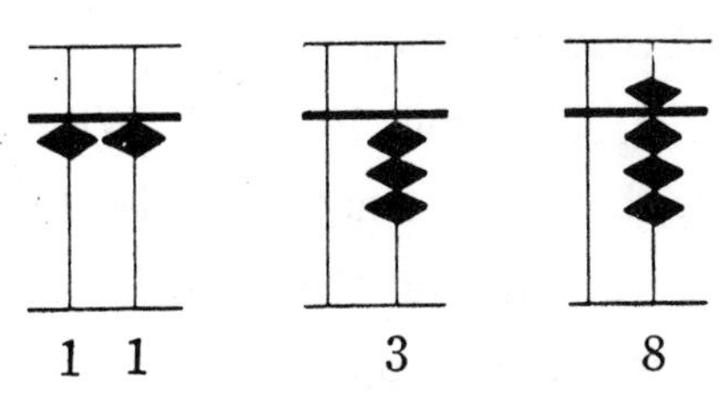

(1) Raise one Earth counter on the 10 upright, with the thumb.

(2) With the thumb also raise one Earth counter on the unit upright. 11 is now indicated on the board.

(3) Think, "S. T. 3 from 10 is 7." With the forefinger return to place the Earth counter on the 10 upright and then press between the fingers the Heaven counter and two Earth counters on the unit upright.

8 is indicated on the board. It is the remainder.

Example 6. 3 from 12 is 9.

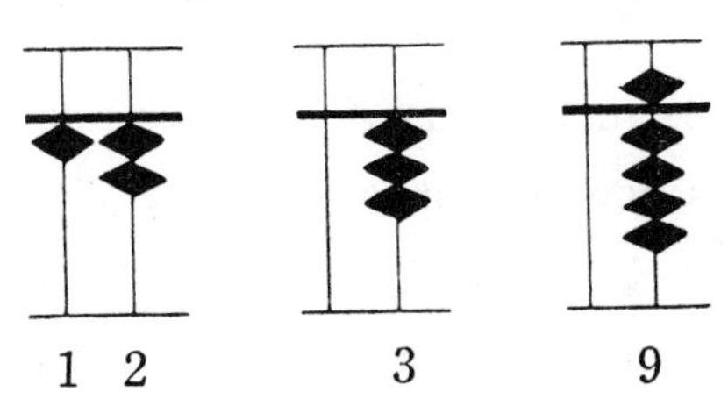

(1) With the thumb raise one Earth counter on the 10 upright.

(2) Raise two Earth counters on the unit upright with the thumb. 12 is now indicated.

(3) Think, "S. T. 3 from 10 is 7." With the forefinger return to place the Earth counter on the 10 upright, then press between the fingers the Heaven counter and two Earth counters on the unit upright.

9 is indicated on the board. It is the remainder.

Example 7. 5 from 14 is 9.

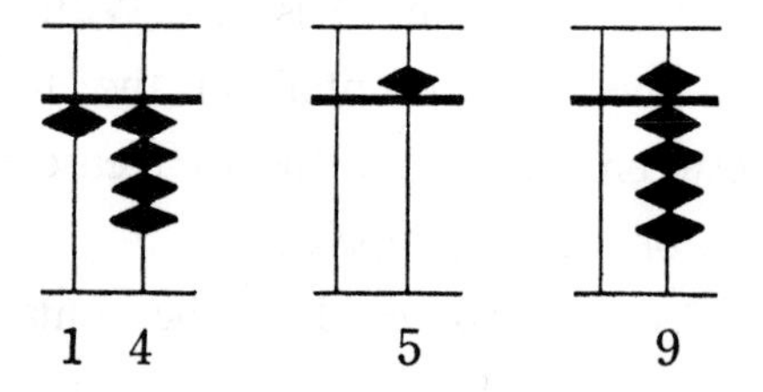

(1) With the thumb raise one Earth counter on the 10 upright.

(2) With the thumb raise four Earth counters on the unit upright. 14 is now indicated.

(3) Think, "S. T. 5 from 10 is 5." Return to place the Earth counter on the 10 upright, using the forefinger and then slide down the Heaven counter on the unit upright.

9 is indicated on the board. It is the remainder.

Example 8. 4 from 11 is 7.

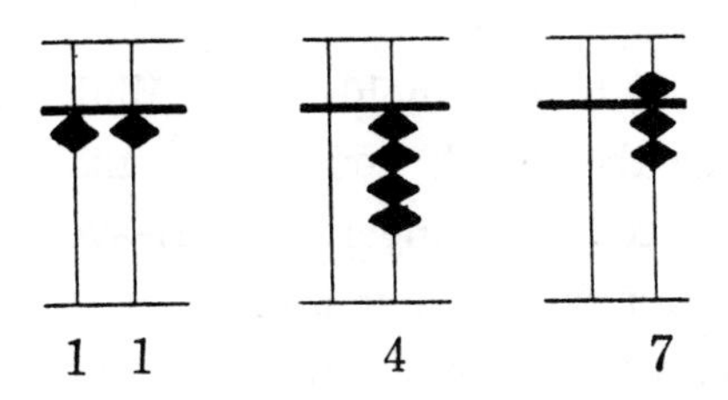

(1) Raise one Earth counter on the 10 upright with the thumb.

(2) With the thumb also, raise one Earth counter on the unit upright. 11 is now indicated.

(3) Think, "S. T. 4 from 10 is 6." With the forefinger return to place the Earth counter on the 10 upright. Then press between the fingers the Heaven counter and one Earth counter on the unit upright.

7 is indicated on the board. It is the remainder.

Example 9. 4 from 13 is 9.

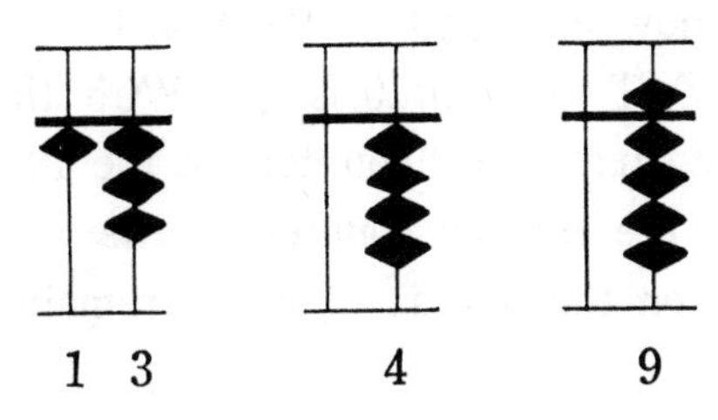

(1) With thumb raise one Earth counter on the 10 upright.

(2) With the thumb raise three Earth counters on the unit upright.

(3) Think, "S. T. 4 from 10 is 6." With the forefinger return to place the Earth counter on the 10 upright and then press between the fingers the Heaven counter, and one Earth counter on the unit upright.

9 is indicated on the board. It is the remainder.

Example 10. 5 from 11 is 6.

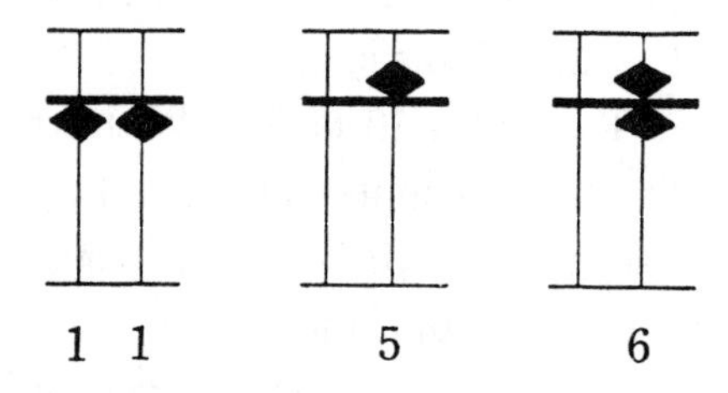

(1) With the thumb raise one Earth counter on the 10 upright.

(2) With the thumb raise one Earth counter on the unit upright. 11 is now indicated on the board.

(3) Think, "S. T. 5 from 10 is 5." With the forefinger return to place the Earth counter on the 10 upright and then slide down the Heaven counter on the unit upright.

6 is indicated on the board. It is the remainder.

Example 11. 6 from 12 is 6.

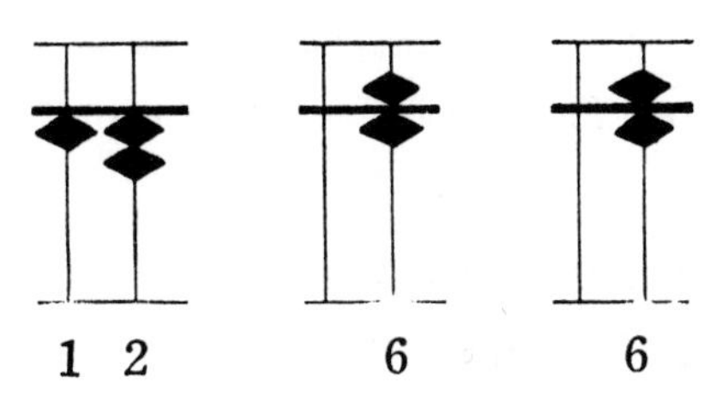

(1) Raise one Earth counter on the ten upright with the thumb.

(2) Raise two Earth counters on the unit upright with the thumb. 12 is now indicated.

(3) Think, "S. T. 6 from 10 is 4," and "A. T. 4 and 1 are 5." Replace the Earth counter on the ten upright with the forefinger and then at one time, slide down the Heaven counter on the unit upright and return one Earth counter on the same upright with the forefinger.

Six is indicated on the board. It is the remainder.

Example 12. 7 from 13 is 6.

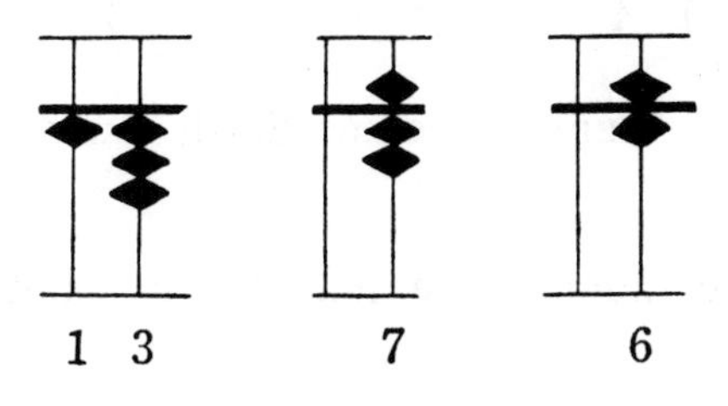

(1) Raise one Earth counter on the ten upright with the thumb.

(2) Raise three Earth counters on the unit upright with the thumb.

13 is now indicated on the board.

(3) Think, "S. T. 7 from 10 is 3 and A. T. 3 and 2 are 5." With the forefinger return to place the Earth counter on the ten upright and then at the same time take down the Heaven counter on the unit upright and return to place two Earth counters on the same upright, with the forefinger.

6 is indicated on the board. It is the remainder.

Example 13. 8 from 14 is 6.

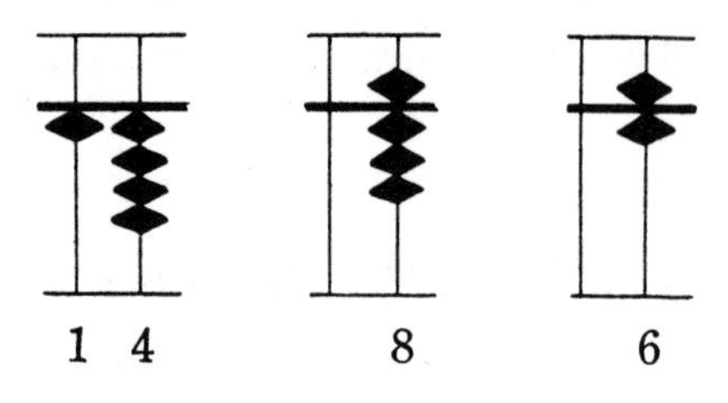

(1) With the thumb raise one Earth counter on the ten upright.

(2) With the thumb raise four Earth counters on the unit upright.

(3) Think, "S. T. 8 from 10 is 2 and A. T. 2 and 3 are 5." With the forefinger return to place the Earth counter on the ten upright. At one time with the forefinger slide down the Heaven counter on the unit upright and return to place three of the Earth counters on the same upright.

6 is indicated on the board. It is the remainder.

Example 14. 9 from 16 is 7.

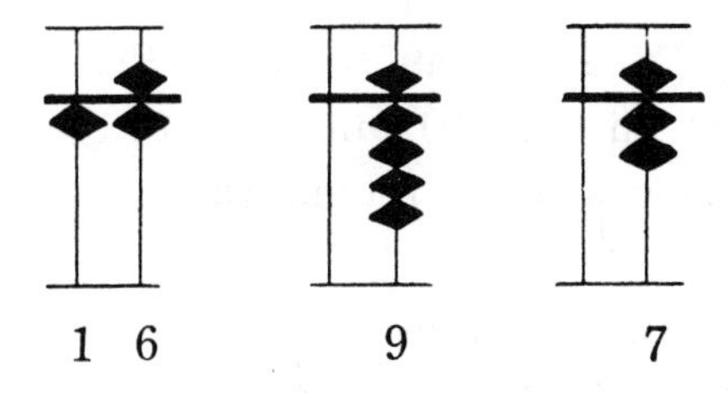

(1) With the thumb raise one Earth counter on the ten upright.

(2) Press between the fingers the Heaven counter and one Earth counter on the unit upright.

16 is now indicated on the board.

(3) Think, "S. T. 9 from 10 is 1." With the forefinger return to place the Earth counter on the ten upright and raise one Earth counter on the unit upright, using the thumb.

7 is indicated on the board. It is the difference.

Example 15. 99 from 100 is 1.

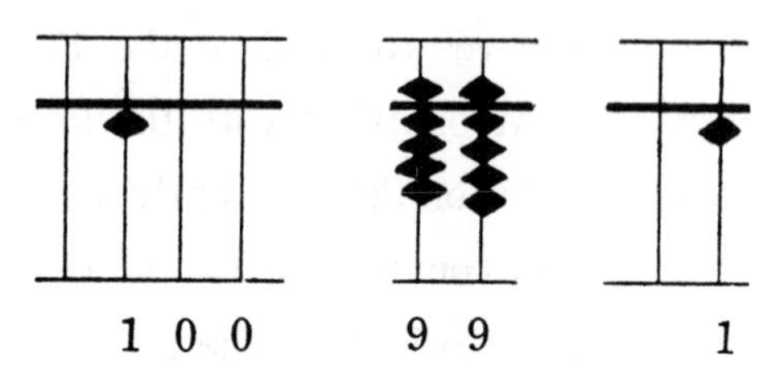

(1) With the thumb raise one Earth counter on the second upright to the left of the unit upright. This indicates 100.

(2) Think, "S. T. 9 from 10 is 1." Return to place the Earth counter on the 100 upright with the forefinger and at the same time with the thumb, raise one Earth counter on the ten upright.

(3) Think, "S. T. 9 from 10 is 1." With the forefinger return to place the Earth counter on the ten upright and then raise one Earth counter on the unit upright.

1 is indicated on the board. It is the remainder.

The above illustrations give practical examples of subtraction with numbers of two places or more. These will show that if you clearly understand subtracting with small numbers, it will be easy for you to handle any number.

In the following exercises find the remainders and write them in position. Check your answers by adding the remainder to the number above it. The sum should be the top number.

Row A.	$ 9.24	$ 5.34	$ 6.85	$ 23.74
	6.68	2.77	3.49	16.88
Row B.	$ 86.03	$ 19.35	$ 24.83	$ 40.52
	76.18	17.84	18.99	24.66

Row C.	$ 2748.36	$ 6025.04	$ 6491.32
	1977.69	4749.27	5943.74
Row D.	$ 1002.30	$ 7341.72	$ 5372.07
	974.64	6488.84	4493.19
Row E.	$ 190.04	$ 43.65	$ 163.28
	111.15	27.89	156.49
Row F.	$ 227.41	$ 395.26	$ 1735.62
	166.83	107.48	1448.83

Subtract and write in the remainders.

Row A.	66 23	66 42	87 43	55 34	44 36	74 46
Row B.	229 154	175 86	408 234	553 378	387 299	
Row C.	15 8	13 9	123 98	897 121	789 667	21 14
Row D.	100 89	1767 989	2325 1936	5566 3179	4858 4075	

Checking Errors in Subtraction.

As was stated before, it is always good to check your answer in subtraction by adding the remainder to the subtrahend, which should give you the minuend.

The following illustrations will make this clear:

67	Minuend	39	Difference
28	Subtrahend	28	Subtrahend
39	Difference	67	Minuend

Section 3.

Subtraction in Cases Where the Subtrahend is Larger than the Minuend.

In such cases we place the subtrahend on the board first and subtract from it the minuend. The result is called the remainder but must be preceded by the minus sign.

Example 1. 7 from 3 is -4.

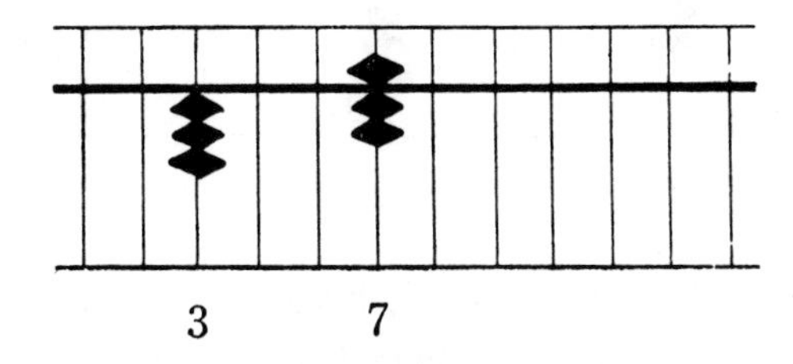

(1) With the thumb raise three Earth counters on an upright in the left half of the abacus.

(2) Press between the fingers one Heaven counter and two Earth counters on an upright in the right half of the abacus.

(3) Because the subtrahend 7 is larger than the minuend 3, we must subtract the minuend from the subtrahend.

Think, "3 from 5 is 2." Raise two Earth counters on

the same upright where 7 is indicated and at the same time return to place the Heaven counter of the same upright.

4 is indicated on the board. It is the remainder if it is preceded by the minus sign.

There is a simpler way to subtract when the subtrahend is larger than the minuend.

Example 2. 9 from 8 is –1.

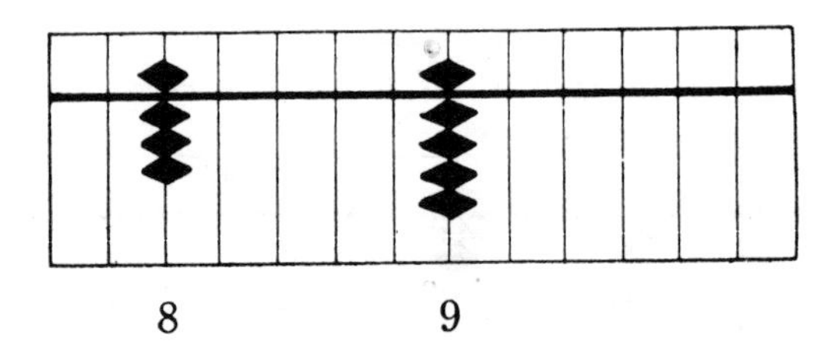

When we subtract 9 from 8, because the 9 is larger than the 8, we can assume that there is one Earth counter raised on the next upright to the left from the unit upright on which 8 is indicated. In other words we assume that we have 18 on the board and then we subtract 9 from 18.

(1) Press between the fingers the Heaven counter and three Earth counters on the unit upright.

(2) Now we are ready to assume that we have ten more than we have, so we raise one Earth counter on the upright to the left of the one where 8 is indicated. 18 is now indicated.

(3) Think, "9 from 10 is 1." With the forefinger return to place the Earth counter on the ten upright and raise one Earth counter on the unit upright, with the thumb. 9 is indicated on the board.

(4) But we have assumed that we have ten more than we really have. Therefore 10 must be subtracted from the 9, leaving us -1.

Whenever we assume that we have 10 or 100 or 1000 more than we really have, we must always subtract this number, which we have assumed, from the remainder, which we get. The difference between these two is the real remainder and is always preceded by a minus sign. It is called the "supplementary number." The following table shows some supplementary numbers:

The supplementary	number of	9	is	1.
,,	,,	8	is	2.
,,	,,	3	is	7.
,,	,,	23	is	77.
,,	,,	567	is	433.
,,	,,	4321	is	5679.

(5) In this case our remainder was 9. The supplementary number of 9 is 1. Therefore, the real remainder is -1.

In the above case we were subtracting with numbers of one place, therefore, we assumed we had ten extra. In subtracting numbers of two places we assume that we have 100 extra, etc.

Find the difference between the lower number and the upper number in the following examples. Lay a strip of paper across the page and write the answers on it. Check your answers with those given.

Practice with these examples until you can do a row or nearly a row in a minute and get them all right.

Row A.	535	430	200	662	239	453
	−278	−192	−135	−483	−171	−335
	257	238	65	179	68	118
Row B.	752	91	87	325	560	97
	−384	−46	−73	−158	−290	−35
	368	45	14	167	270	62
Row C.	357	528	812	200	998	677
	−159	−379	−576	−157	−372	−257
	198	149	236	43	626	420

Write the difference, making sure that each remainder is right by adding it to the subtrahend.

Row D.	$6.74	$9.03	$7.59	$4.59
	−3.21	−4.09	−6.52	−2.39

Review.

Addition and Subtraction in combination.

Give as many right answers as you can in a minute.

(1)	(2)	(3)	(4)	(5)
$ 24.72	$ 1.52	$ 222.33	$ 51.82	$ 45.58
12.15	- .73	-34.54	671.35	6.63
-9.98	- .32	- 9.93	78.56	112.96
3.21	13.54	-18.97	-442.84	- 10.07
17.05	52.53	-66.26	- 77.15	- 55.99
75.37	5.51	-31.74	- 19.91	101.20
-21.12	-20.03	-13.80	627.84	70.81
4.51	7.94	-29.17	-791.17	22.56
22.37	13.49	- 8.33	- 84.91	18.73
-44.78	4.44	- 6.66	44.44	-203.52
- 7.11	-17.19	2.52	21.47	- 39.99
18.36	- 3.24	4.65	377.99	131.10
7.71	41.82	17.92	222.35	- 73.75
18.01	15.92	1.88	- 97.14	27.75
1.01	16.99	10.10	- 24.58	18.49
-91.38	- 75.91	4.57	121.43	13.18
40.51	34.88	5.35	240.52	526.17
-21.22	- 42.27	- 14.43	99.13	288.04
-35.53	- 19.99	- 18.19	- 27.32	-444.44
6.16	- 18.90	2.02	- 18.62	-189.44
$ 20.02	$ 10.00	$ 19.32	$973.26	$366.00

Balancing of

Receipts and Expenditures.

Date	Receipts	Expenditures	+ or −	Balance
1	$ 5762.61	$ 3473.49	+	$ 2289.12
2	937.54	658.35	+	279.19
3	48.15	239.98	−	191.83
4	619.50	422.60	+	196.90
5	2544.32	1945.73	+	598.59
6	91.02	877.41	−	786.39
7	793.67	682.85	+	110.82
8	214.86	122.40	+	92.46
9	6022.33	4751.45	+	1270.88
10	214.86	122.40	+	92.46
11	85.27	471.37	−	386.10
12	151.93	99.16	+	52.77
13	830.24	561.58	+	268.66
14	2481.10	3354.37	−	873.27
15	65.83	29.41	+	36.42
16	229.48	198.46	+	31.02
17	5216 77	6627.18	−	1410.41
18	889.63	792.53	+	97.10
19	510.33	798.46	−	288.13
20	78.49	66.82	+	11.67
21	114.85	203.55	−	88.70
22	862.00	750.43	+	111.57
23	1284.56	2114.85	−	830.29
24	772.09	976.48	−	204.39
25	3740.10	2963.38	+	776.72
26	101.66	310.55	−	208.89
27	563.28	93.62	+	469.66
28	1529.47	1003.41	+	526.06
29	209.53	2133.46	−	1923.93
30	773.37	471.20	+	302.17
	37738.84	37316.93		421.91

CHAPTER IV.

MULTIPLICATION

There are two methods of multiplication on the abacus in general use at the present time. By one method we begin to multiply the last figures in each of the numbers together, but in the other method we begin by multiplying the first figure of one number with the last figure of the other.

The first of these methods is called, "Multiplication Beginning at the Right."

The second of these methods is called, "Multiplication Beginning at the Left."

18	Multiplicand
9	Multiplier
162	Product

In "Multiplication Beginning at the Right" we begin the multiplication by multiplying the last figure of the multiplier and the last figure of the multiplicand and continue working to the left.

In "Multiplication Beginning at the Left" we begin our multiplication by multiplying the first figure of the multiplier and the last figure of the multiplicand, working to the left in the multiplicand and to the right in the multiplier.

The first of these methods, "Multiplication Beginning at the Right" is convenient because it is already familiar to us in ordinary multiplication, but the other method, "Multiplication Beginning at the Left", is really more convenient, if by practice we become skilful in it.

By explanation I shall try to show you both of these methods.

Section 1.

Multiplication Table.

In this section the abbreviation "M. T." stands for Multiplication Table. It is very important for you to memorize this table.

Multiplication Table.

M. T. of 1.				
1	×	1	is	1.
1	×	2	is	2.
1	×	3	is	3.
1	×	4	is	4.
1	×	5	is	5.
1	×	6	is	6.
1	×	7	is	7.
1	×	8	is	8.
1	×	9	is	9.

M. T. of 2.				
2	×	2	is	4.
2	×	3	is	6.
2	×	4	is	8.
2	×	5	is	10.
2	×	6	is	12.
2	×	7	is	14.
2	×	8	is	16.
2	×	9	is	18.

M. T. of 3.

3	×	3	is	9.
3	×	4	is	12.
3	×	5	is	15.
3	×	6	is	18.
3	×	7	is	21.
3	×	8	is	24.
3	×	9	is	27.

M. T. of 4.

4	×	4	is	16.
4	×	5	is	20.
4	×	6	is	24.
4	×	7	is	28.
4	×	8	is	32.
4	×	9	is	36.

M. T. of 5.

5	×	5	is	25.
5	×	6	is	30.
5	×	7	is	35.
5	×	8	is	40.
5	×	9	is	45.

M. T. of 6.

6	×	6	is	36.
6	×	7	is	42.
6	×	8	is	48.
6	×	9	is	54.

M. T. of 7.

7	×	7	is	49.
7	×	8	is	56.
7	×	9	is	63.

M. T. of 8.

8	×	8	is	64.
8	×	9	is	72.

M. T. of 9.

9	×	9	is	81.

When you multiply two numbers, you must constantly keep in mind the Multiplication Table. And sometimes it is even convenient to say it out loud while manipulating the counters.

Section 2.

Determination of Position in Multiplication.

When we multiply two numbers we set the multiplier on the lefthand side and the multiplicand to the right of it. Then we must determine the unit position of the product.

Count the number of uprights in the multiplier. Count off the same number of uprights from the multiplicand towards the right. The last one is the unit position for the product.

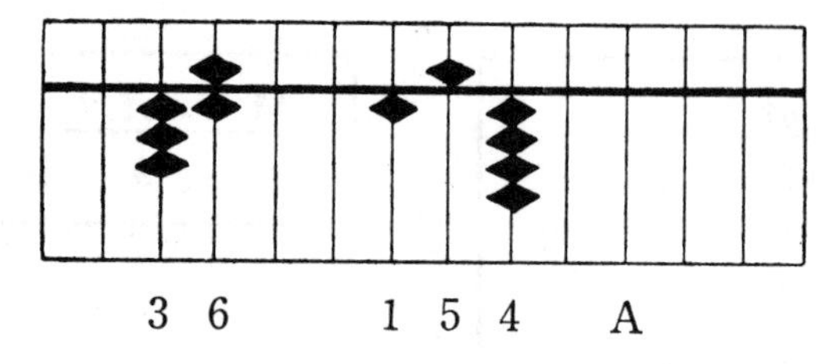

In the above illustration the A upright is the unit upright of the product.

36 multiplier
154 multiplicand

In this case two uprights are necessary to indicate the multiplier. Therefore, we count off two uprights from the multiplicand, and the second one is the unit position of the product.

Section 3.

Multiplication Beginning at the Right.

This is a method of multiplication which prevails at the present time.

When we begin to multiply by this method, we first multiply the righthand figure of the multiplier by the right-hand figure of the multiplicand, gradually working to the left.

Part 1.

The Process of Multiplication.

(1) We must indicate the multiplicand on the righthand side of the board, leaving space for the product. And we must indicate the multiplier on the lefthand side of the board.

(2) As was stated before we must count the number of uprights in the multiplier, then count off an equal number of uprights to the right of the multiplicand.

The last one is the unit position for the product.

Now you are ready to begin the multiplication.

(1) Multiply the end figure of the multiplicand by the end figure of the multiplier according to the Multiplication Table.

(2) If we get a product of two places such as 32, we must indicate the two on the unit position of the product by raising two Earth counters, and indicate the three on the upright to the left of the unit one by raising three Earth counters.

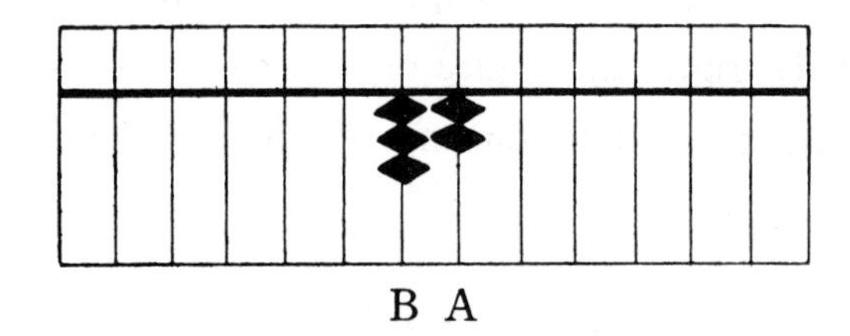

A is the upright for the unit position

B is the upright for 10's

(3) Multiply the figure on the ten upright of the multiplier and the end figure of the multiplicand, using the Multiplication Table.

(4) If we should get a product of one place such as 6, we must indicate the 6, on the ten position of the product.

(5) If we should get a product of two places, such as 45, we must indicate the 5 on the ten upright and the four on the hundred upright of the product as in the following:

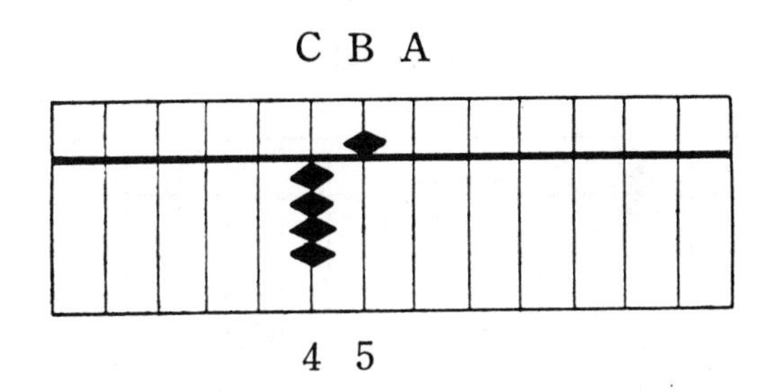

A is the upright for ones
B is the upright for tens
C is the upright for hundreds

Continuing like this we will come to the place where we will have multiplied the end figure of the multiplicand by every figure of the multiplier, then we must begin to multiply the next figure of the multiplicand by each figure of the multiplier.

When we have multiplied every figure of the multiplicand by every figure of the multiplier the product is indicated on the board.

Part 2.

Example 1.

56	multiplicand
6	multiplier
336	product

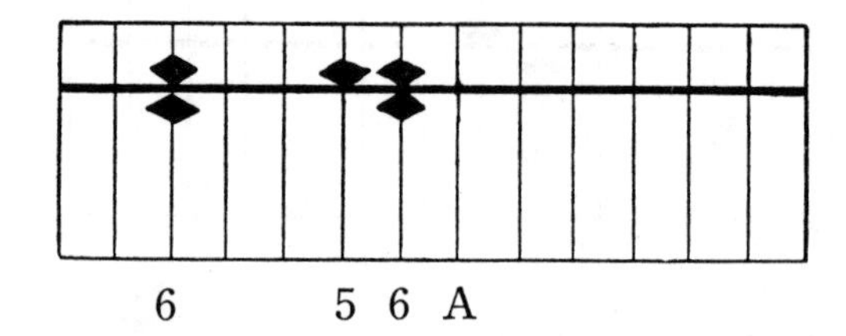

In this case A is the unit upright of the product.

(1) Indicate the multiplier on the lefthand side of the board and the multiplicand to the right of it, as shown in the above illustration.

(2) There is one upright in the multiplier, therefore, the first upright to the right of the multiplicand is the unit upright of the product.

(3) Multiplying the end figures of the multiplier and the multiplicand we get, "6×6 is 36."

(4) Return to place the Heaven counter and the Earth counter on the unit position of the multiplicand, and then

raise three Earth counters on this upright, which is the position for tens in the product, and press between the fingers the Heaven counter and one Earth counter on the next upright to the right, which is the unit position of the product. 536 is now indicated.

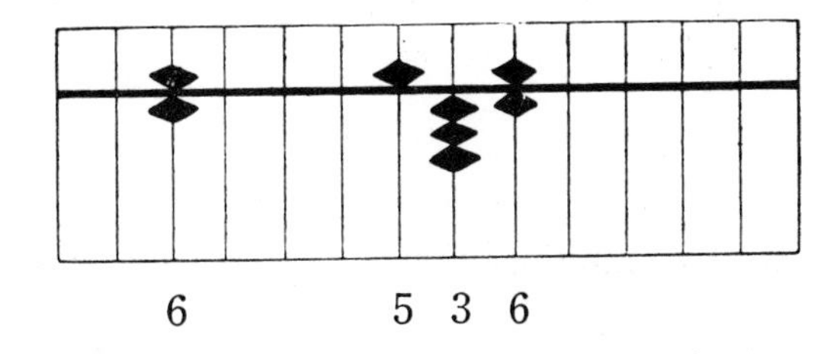

(5) As there is only one figure in the multiplier we are now ready to multiply the multiplier and the second figure of the multiplicand, which is 5. "6×5 is 30."

(6) Return the Heaven counter on the first upright of the multiplicand and raise three Earth counters on the same upright, as it is the upright for 100's in the product.

336 is indicated on the board. It is the product.

Example 2. 56 multiplicand
12 multiplier

672 product

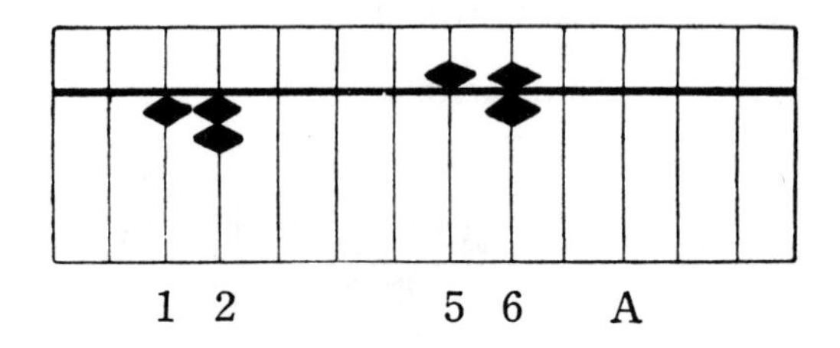

(1) Indicate the multiplier on the lefthand side of the board and the multiplicand on the righthand side as shown in the above illustration.

(2) There are two uprights in the multiplier, therefore, count off two uprights to the right of the multiplicand to find the unit position of the product.

(3) Multiply the end figure of the multiplicand by the end figure of the multiplier, "6×2 is 12."

(4) Raise one Earth counter on the upright for tens in the product and raise two Earth counters on the unit upright of the product. 5612 is indicated on the board.

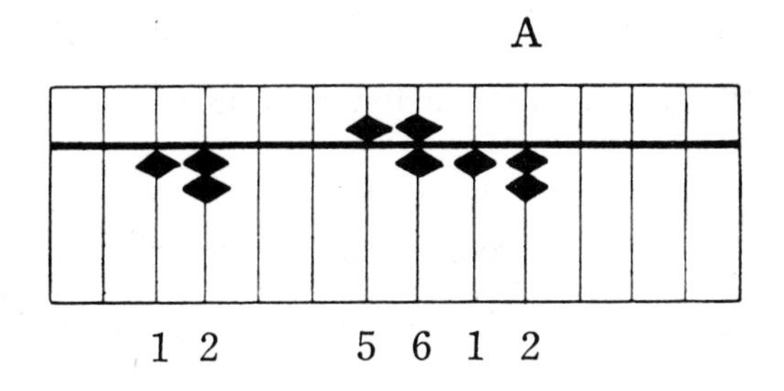

(5) Next multiply the 1 of the multiplier and the 6 of the multiplicand. This gives us "1×6 is 6."

(6) Because we have now multiplied each figure of the multiplier by the figure 6 of the multiplicand, we must now return these counters to place. But we have not yet indicated on the board the product 6 which we got by multiplying the first figure of the multiplier by the last figure of the multiplicand. We now indicate this 6 on the upright for 10's of the product. 5072 should now be indicated on the board.

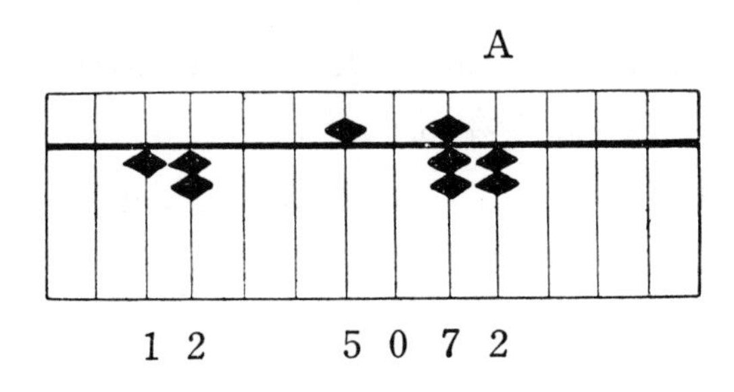

(7) We are now ready to multiply the first figure of the multiplicand, 5, with the end figure of the multiplier. "5×2 is 10."

(8) Raise one Earth counter on the upright for 100's in the product. 5172 should now be indicated on the board.

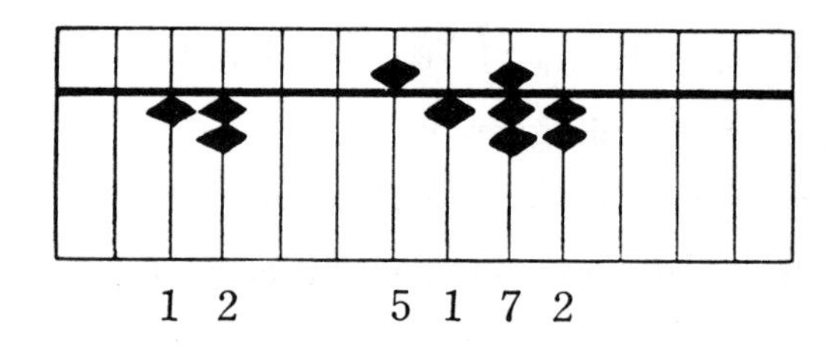

(9) We now multiply the 5 of the multiplicand and the figure 1 of the multiplier. "5×1 is 5."

(10) Because we have multiplied the figure 5 of the multiplicand by each figure of the multiplier we now return to place this 5. Then we must indicate the product 5 which we have just gotten. It must be placed on the upright for 100's in the product.

672 is now indicated on the board. It is the product.

Example 3. 142 multiplicand
95 multiplier
13490 product

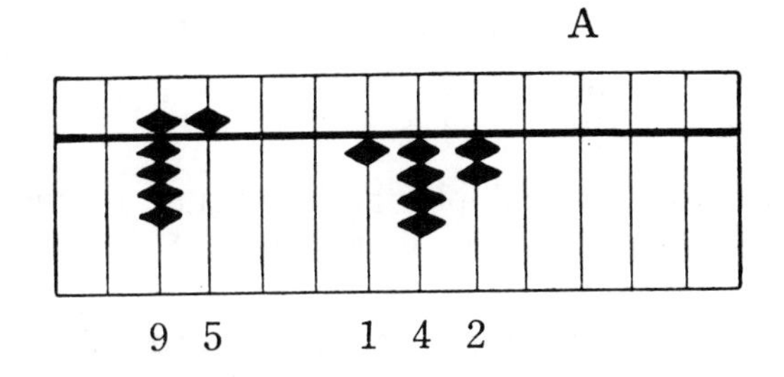

(1) Indicate the multiplier 95 on the lefthand side of the board and the multiplicand 142 on the right of it, as is shown in the above illustration.

(2) As there are two uprights in the multiplier, count off two places to the right of the multiplicand. This determines the position for the units of the product.

(3) Multiply the end figure of the multiplicand and the end figure of the multiplier. "2×5 is 10."

(4) Raise one Earth counter on the 10 upright of the product, and 14210 will be indicated on the board.

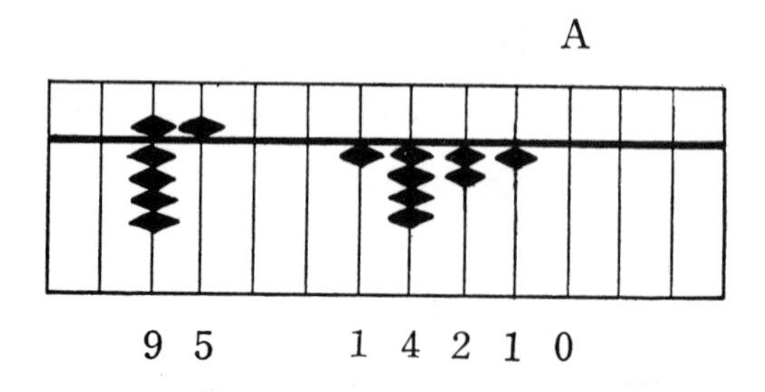

(5) We must now multiply the last figure of the multiplicand which is 2, by the first figure of the multiplier, which is 9. "2×9 is 18."

(6) Because we have multiplied the end figure of the multiplicand by every figure of the multiplier, we are now ready to return to place the counters which indicate this end figure, which is 2. But we can see that the 1 of the product 18 must be indicated on the same upright with the 2 which we are ready to erase, so we return to place only 1 of the Earth counters, leaving 1 to indicate the 1 of 18. Then on the next upright to the right, we indicate the 8 by pressing between our fingers 1 Heaven counter and three Earth counters. 14190 should now be indicated on the board.

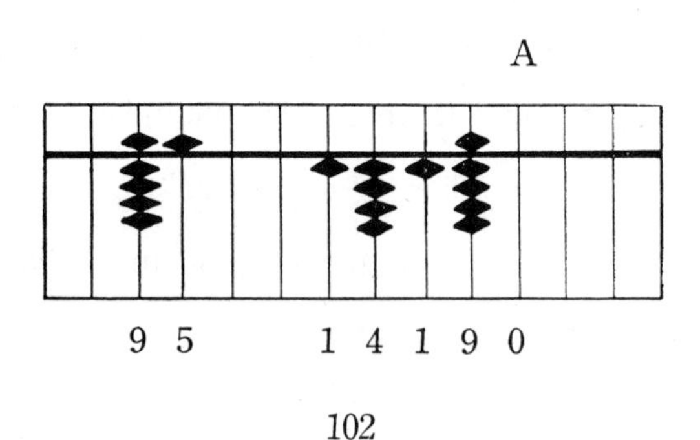

(7) We are now ready to multiply the second figure of the multiplicand by the multiplier. Multiplying the end figure of the multiplier with the second figure of the multi-pilicand we get, "4×5 is 20."

(8) Raise two Earth counters on the upright for 100's in the product. 14390 should now be indicated.

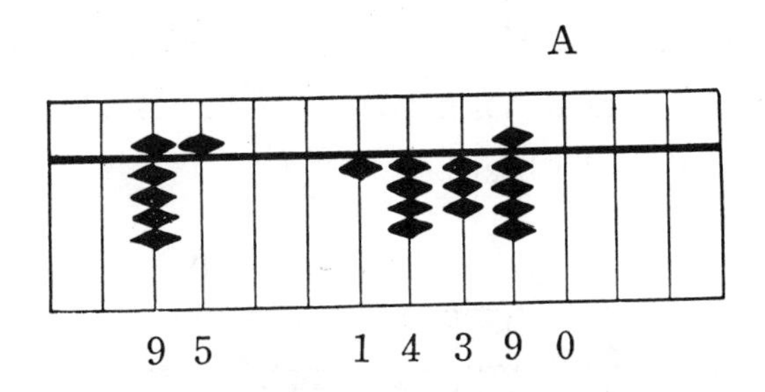

(9) The second figure of the multiplicand must now be multiplied by the first figure of the multiplier, that is, "4×9 is 36."

(10) Instead of returning all four of the counters in the figure 4 which we no longer need, we take away only one, leaving three to indicate the 30 of 36, and then indicate 6 on the next upright to the right. 13990 should now be indicated on the board.

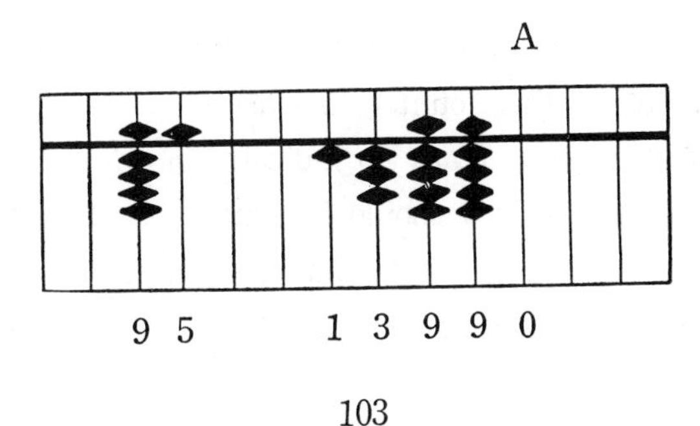

(11) Our next step is to begin to multiply the first figure of the multiplicand by the end figure of the multiplier. "1×5 is 5."

(12) We wish to indicate this 5 on the upright for 100's in the product but 9 is indicated already. Therefore, we return to place the Heaven counter and raise one Earth counter on the next upright to the left. 14490 should now be indicated on the board.

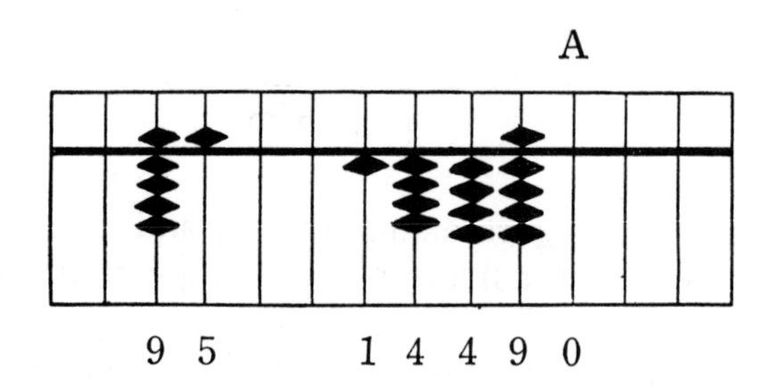

(13) We must now multiply the 1, which is the first figure of the multiplicand, and 9 which is the first figure of the multiplier. "9×1 is 9."

(14) As we have now finished with the 1 of the multiplicand, we return this counter to place.

(15) We are now ready to indicate the product which we got in step 13. This 9 should be placed on the upright for 1000's in the product, but as that upright is already almost full it is easier to replace one Earth counter on the upright

for 1000's and at the same time raise one Earth counter on the next upright to the left.

13490 is indicated on the board. It is the product.

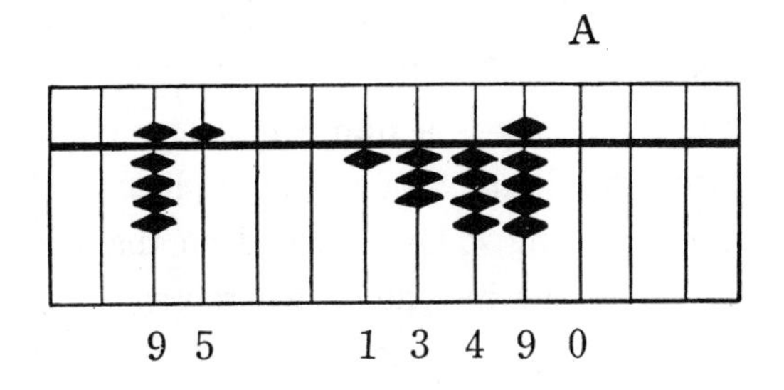

Part 3.

Examples for Practice.

Exercise 1.

53	multiplicand
7	multiplier
371	product

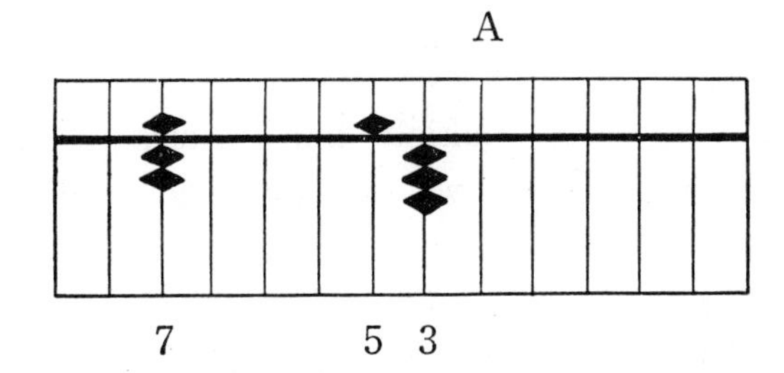

(1) Indicate the multiplier 7 on the lefthand side of the abacus, and to the right of it, the multiplicand 53.

(2) As there is only one upright in the multiplier, the first upright on the right of the multiplicand becomes the unit position of the product.

(3) Multiplying the end figure of the multiplicand by the multiplier we get, "3×7 is 21."

(4) As the multiplier is a number of one place only, we no longer need the figure 3 of the multiplicand. But as the 2 of our product 21 must be indicated on the same upright on which the 3 is now indicated, we return to place only one of these Earth counters, leaving two to indicate the 20 of 21. We then raise one Earth counter on the unit upright. 521 should now be indicated on the board.

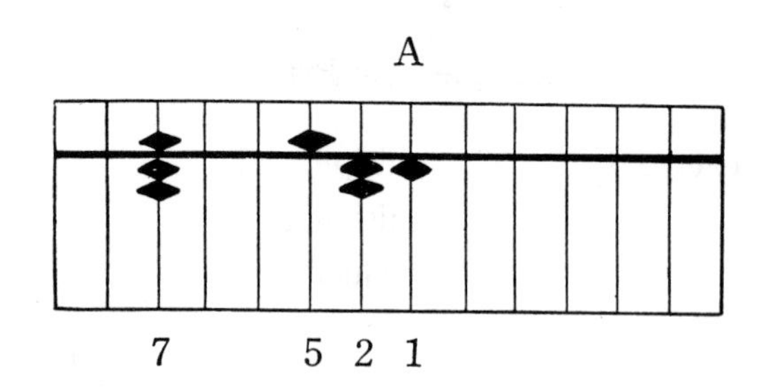

(5) We must now multiply the 5 which is the second figure of the multiplicand with the multiplier. This gives us "5×7 is 35."

(6) Return to place the figure 5 of the multiplicand, as the multiplier is a one place number and we have no further use for the 5. Raise three Earth counters on this same upright as it is the upright for 100's in the product. Then

slide down the 5 counter of the next upright to the right, the position for 10's in the product.

371 is indicated on the board. It is the product.

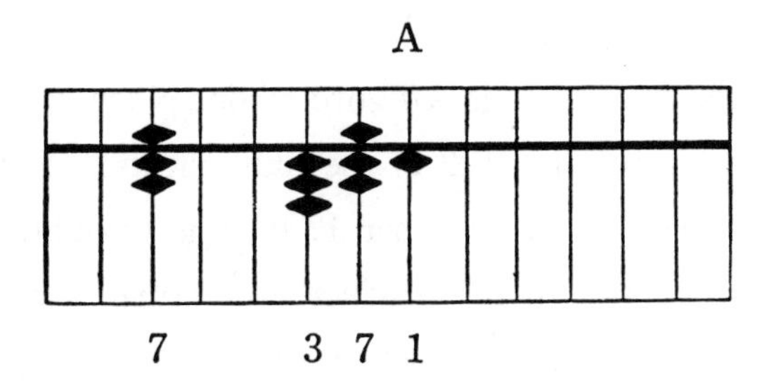

Exercise 2.

232	multiplicand
6	multiplier
1392	product

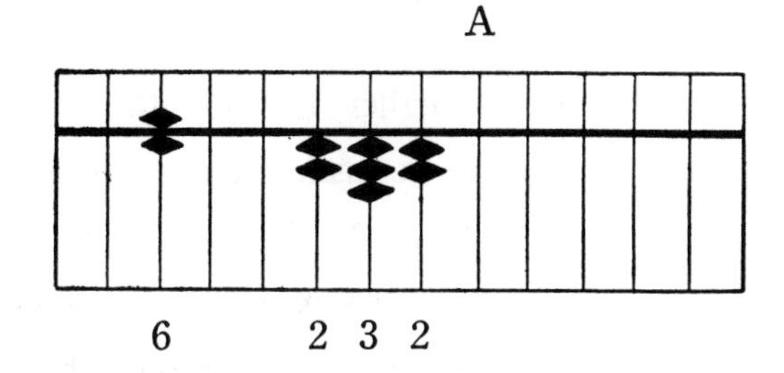

(1) On the lefthand side of the abacus indicate the multiplier 6, and to the right of it the multiplicand 232, as is shown in the above illustration.

(2) As there is only one upright in the multiplier, the

next upright to the right of the multiplicand is the unit position for the product.

(3) Multiplying the end figure of the multiplicand by the multiplier, we get, "2×6 is 12."

(4) As the multiplier is a figure of one place, we have no further use for the figure 2 of the multiplicand. But as we must place the 1 of the number 12 on this upright, as it is the position for 10's in the product, we return to place only one of the two counters. Then we raise two Earth counters on the unit upright of the product. 2312 should now be indicated on the board.

(5) We are now ready to multiply the second figure of the multiplicand by the multiplier. This gives us, "3×6 is 18."

(6) We now have no further use for the figure 3 of the multiplicand, but before returning these three counters to place we remember that the 1 of 18 must be indicated on that same upright, as it is the position for 100's in the product. Therefore, we return to place only two of the three Earth counters. Then we press between the fingers the Heaven counter and three Earth counters on the next upright to the right, or the upright for 10's in the product. 2192 is now indicated on the board.

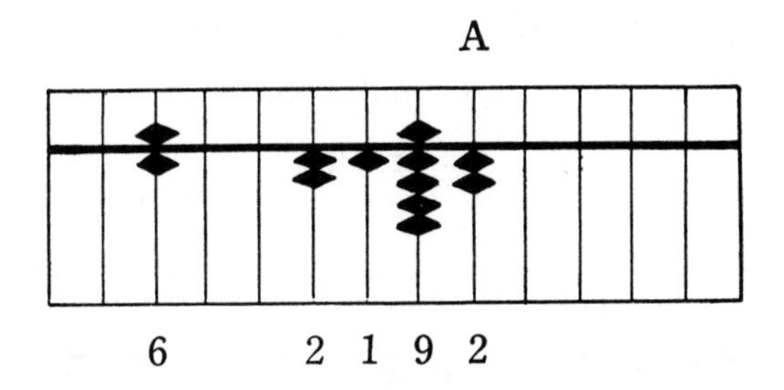

(7) We are now ready to multiply the one remaining figure of the multiplicand by the multiplier. This gives us, "2×6 is 12."

(8) We no longer need the figure 2 indicated on the board, but as we must indicate the 1 of 12 on the same upright we return to place only one of the two counters. Then we raise two Earth counters on the next upright to the right to indicate the 2 of 12.

1392 is indicated on the board. It is the product.

In the United States one dollar ($1.00) is equal to one hundred cents (100¢). When multiplying with figures representing money, do not think of the decimal point until you have gotten the product. Then place the decimal point between the position for 10's and for 100's. This will give you the figure in dollars and cents.

Exercise 3. $\quad$ \$ 5.13 multiplicand

.23 multiplier

\$ 117.99 product

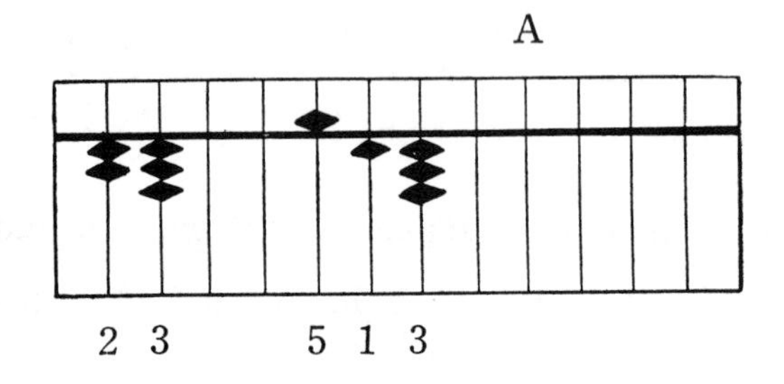

(1) Indicate the multiplier 23 on the lefthand side of the board. To the right of it indicate the multiplicand 513, as shown in the above illustration.

(2) As there are two uprights in the multiplier, the second upright from the multiplicand to the right, is the upright for the unit position of the product.

(3) The end figure of the multiplicand multiplied by the end figure of the multiplier gives us, "3×3 is 9."

(4) Press between the fingers the Heaven counter and four Earth counters on the unit position of the product. 51309 is indicated on the board.

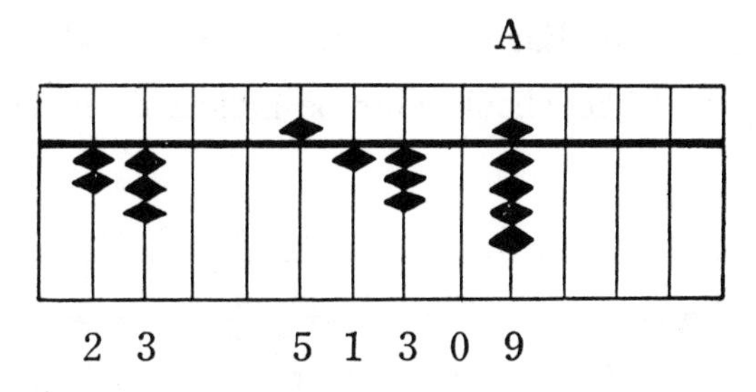

(5) As the multiplier is a two place number we must next multiply the end figure of the multiplicand by the second figure of the multiplier. This gives us, "3×2 is 6."

(6) We have now finished with the figure 3 of the multiplicand, so we return these three counters to place. Then we press between our fingers the Heaven counter and one Earth counter on the upright for 10's in the product. 51069 is now indicated on the board.

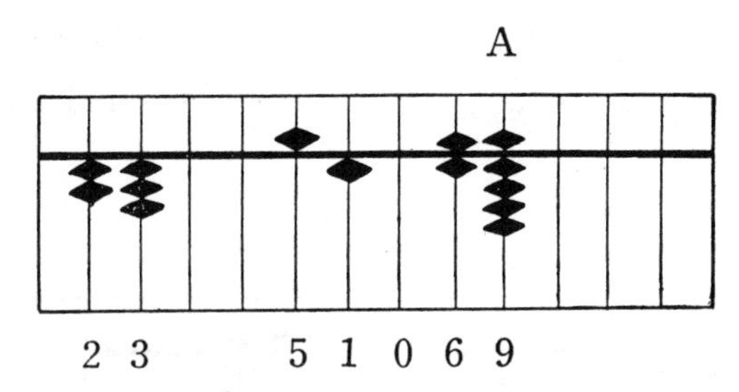

(7) We are now ready to multiply the second figure of the multiplicand with the end figure of the multiplier. This gives us, "1×3 is 3."

(8) Raise three Earth counters on the upright for 10's in the product. 51099 should now be indicated on the board.

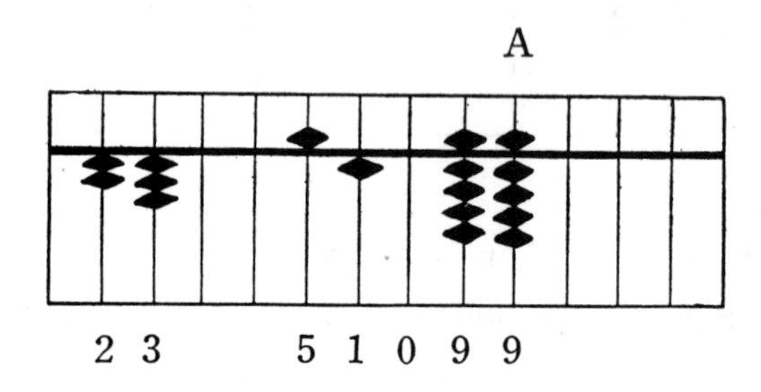

(9) Next we must multiply the 1 of the multiplicand with the second figure of the multiplier. This gives us, "1×2 is 2."

(10) As we have no further use for the one of the multiplicand, we return it to place and raise two Earth counters on the next upright to the right, which is the place for 100's in the product. 50299 should be indicated on the board.

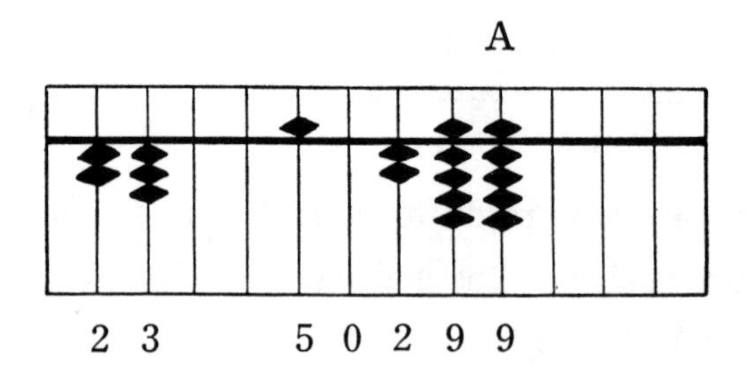

(11) Multiplying the first figure of the multiplicand by the end figure of the multiplier we get, "5×3 is 15."

(12) Raise one Earth counter on the upright for 1000's in the product and slide down the Heaven counter on the upright for 100's in the product. 51799 should be indicated on the board.

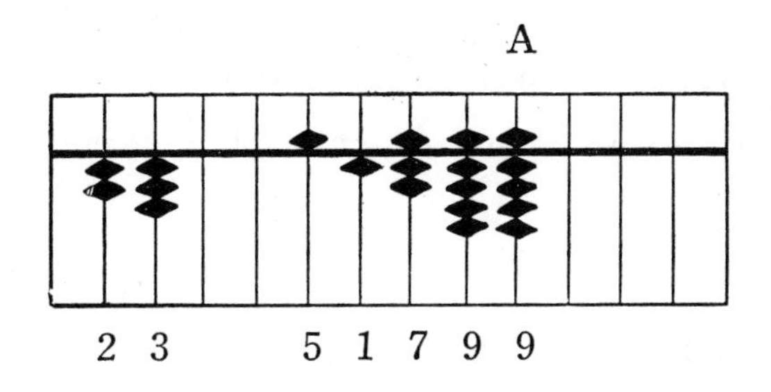

(13) Multiplying the first figure of the multiplicand by the first figure of the multiplier we get, "5×2 is 10."

(14) As we now have no further use for the first figure of the multiplicand, we must return it to place. Next we raise one Earth counter on the upright for 10,000's in the product to indicate the 10 which we got in step 13.

11799 is now indicated on the board. This is the product.

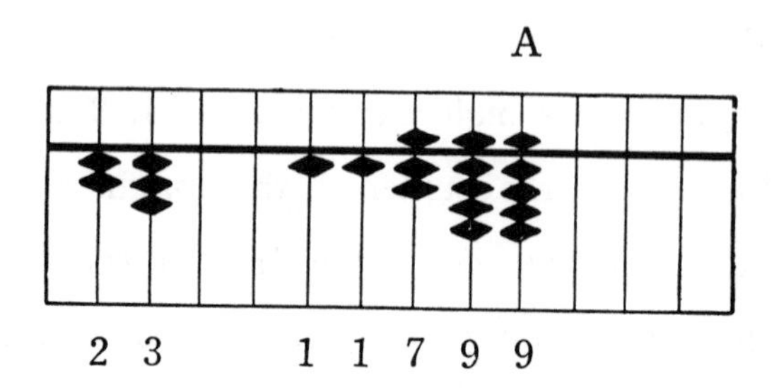

Now we must place a decimal point between the upright for 10's and the upright for 100's in the product.

The product will then be indicated as $117.99.

In just this way, you can multiply any numbers large or small.

Exercise 4. 2519 multiplicand
43 multiplier
108317 product

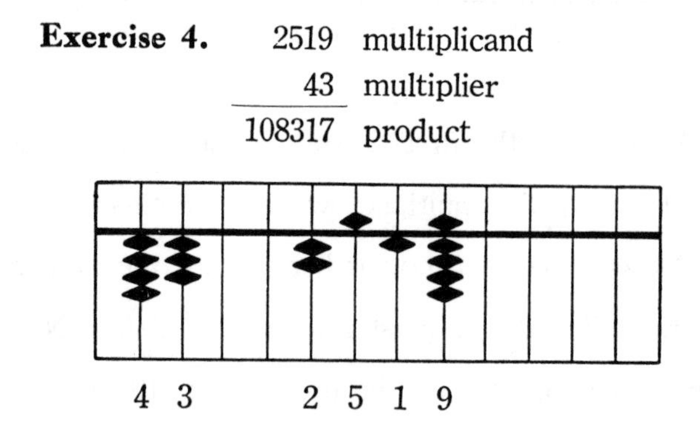

(1) Indicate the multiplier 43 on the lefthand side of the board, and to the right of it the multiplicand, 2519, as is shown in the above illustration.

(2) As there are two uprights in the multiplier, the second upright to the right of the multiplicand is the unit upright for the product.

(3) Multiplying the end figure of the multiplicand with the end figure of the multiplier we get, "3×9 is 27."

(4) Raise two Earth counters on the upright for 10's in the product and press between the fingers the Heaven counter and two Earth counters on the unit upright of the product. 251927 should now be indicated on the board.

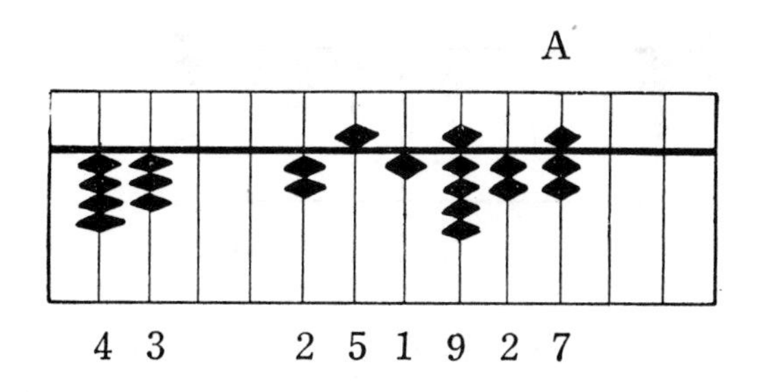

(5) Multiplying the end figure of the multiplicand by the first figure of the multiplier, we get, "9×4 is 36."

(6) We should now return to place the counters which indicate 9, the end figure of the multiplicand, as we have no further use for it. But we can see that the figure 3 of

the product 36 must be indicated on the same upright, so we return to place the Heaven counter and one Earth counter, leaving 3 indicated on the upright for 100's in the product. Then we press between our fingers the Heaven counter and one Earth counter on the next upright to the right, which places our 6 on the upright for 10's in the product. The board should now indicate 251387.

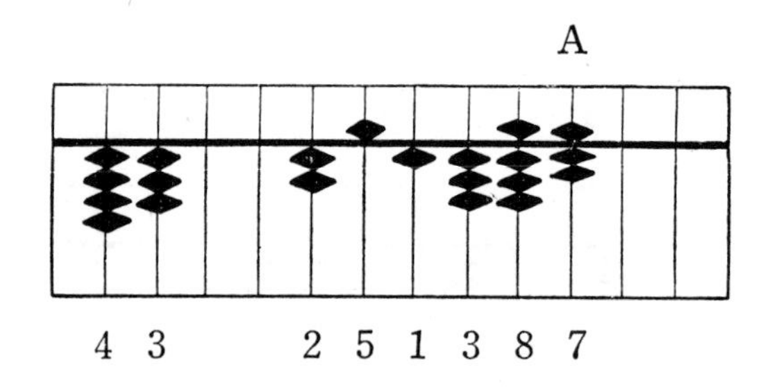

(7) Multiplying the next figure of the multiplicand with the end figure of the multiplier, we get, "1×3 is 3."

(8) We wish to raise three Earth counters on the upright for 10's in the product, but from the previous multiplication 8 is already indicated on that upright. Therefore, we think, "A. T. 3 and 7 are 10," and we return to place the Heaven

counter and two Earth counters on that upright, at the same time raising one Earth counter on the next upright to the left. The board should now indicate 251417.

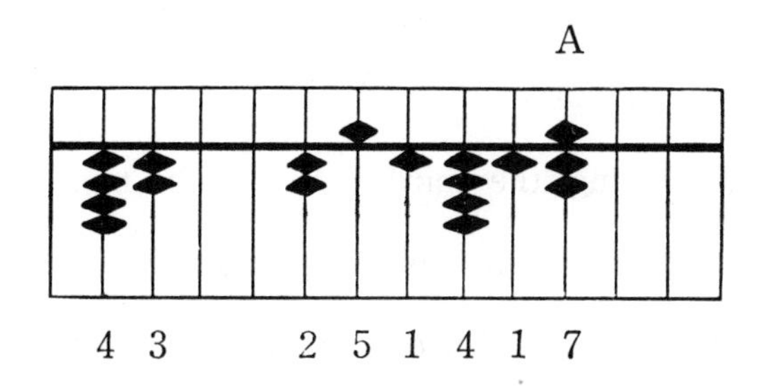

(9) Multiplying this same figure of the multiplicand, 1, by the first figure of the multiplier, we get, "1×4 is 4."

(10) Before indicating this product we must return to place this figure as we have now completed our multiplications. We now want to indicate 4 on the upright for 100's in the product, but there are already four Earth counters in position on that upright, so we think, "A. T. 4 and 1 are 5." We slide down the Heaven counter and at the same time return one Earth counter. The board should now indicate 250817.

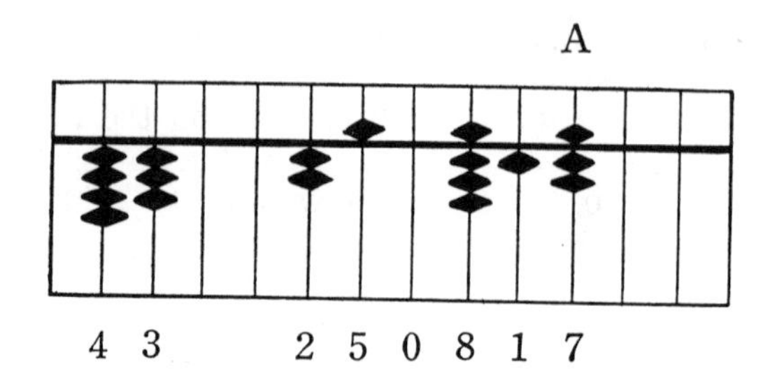

(11) Multiplying the third figure of the multiplicand by the end figure of the multiplier, we get, "3×5 is 15."

(12) Raise one Earth counter on the upright for 1000's in the product. The upright for 100's in the product, on which we wish to indicate 5 has 8 already indicated on it. Therefore, we must think, "A. T. 5 and 5 are 10," and return to place the Heaven counter on that upright, at the same time raising one Earth counter on the next upright to the left.

The board should now indicate 252317.

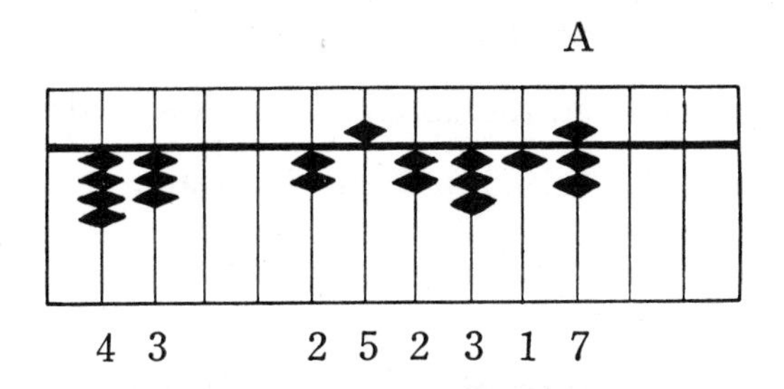

(13) The third figure of the multiplicand multiplied by the first figure of the multiplier gives us, "5×4 is 20."

(14) As we now have no further use for this third figure of the multiplicand we return it to place and then record the multiplication by raising two Earth counters on the upright for 10,000's in the product. 222317 should now be indicated on the board.

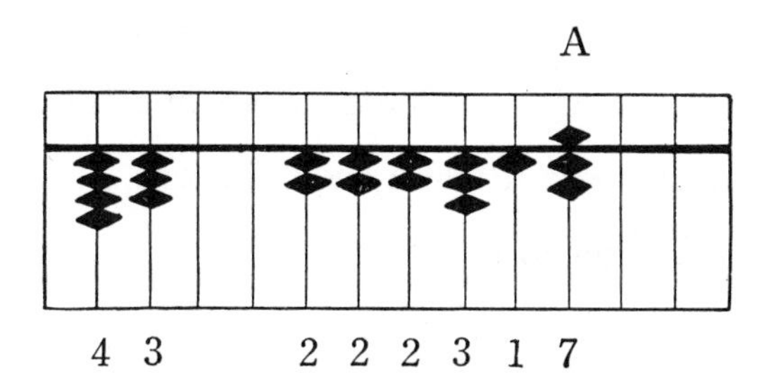

(15) The first figure of the multiplicand multiplied by the end figure of the multiplier gives us, "2×3 is 6."

(16) Press between the fingers the Heaven counter and one Earth counter on the upright for 1000's in the product. 228317 should now be indicated on the board.

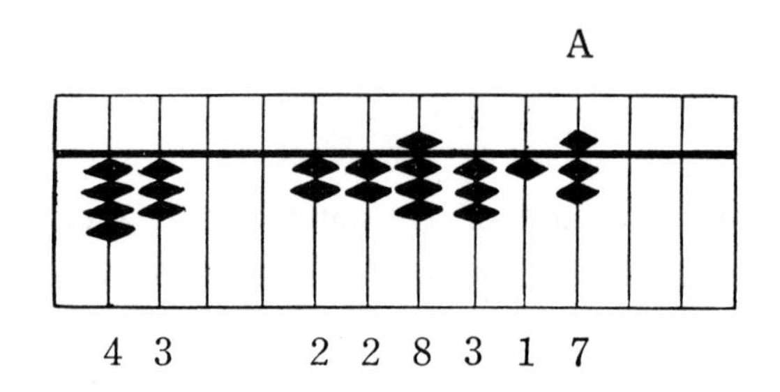

(17) Multiplying the first figure of the multiplicand by the first figure of the multiplier, we get, "2×4 is 8."

(18) We have no further use for the figure 2 of the multiplicand, so we return it to place. The upright for 10,000's in the product upon which we wish to record the product 8 has two Earth counters already in position. Therefore, we return these to place and raise one Earth counter on the next upright to the left.

We have now reached the final product. The board should indicate 108317.

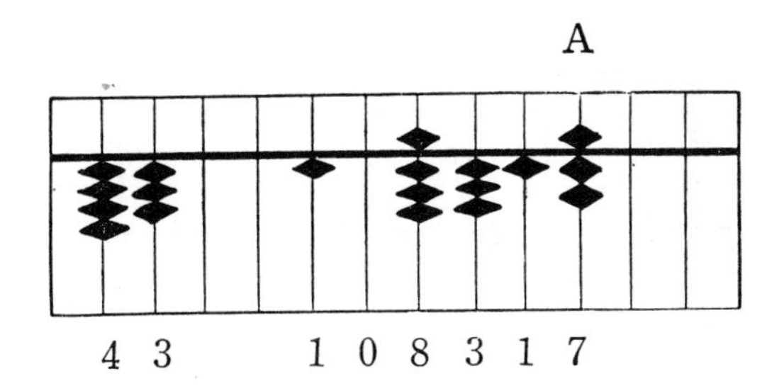

Having completed these exercises we should understand a little of the method of "Multiplication Beginning at the Right." But in case you do not thoroughly understand, I wish to try again to make this method more clear.

When the multiplier is a one place number, the unit position of the product is the next upright to the right of the unit upright of the multiplicand.

We must always decide the unit position of the product first. In long numbers if we find a unit position with each multiplication, the position will advance one upright to the left each time.

I will show you by example what I mean:

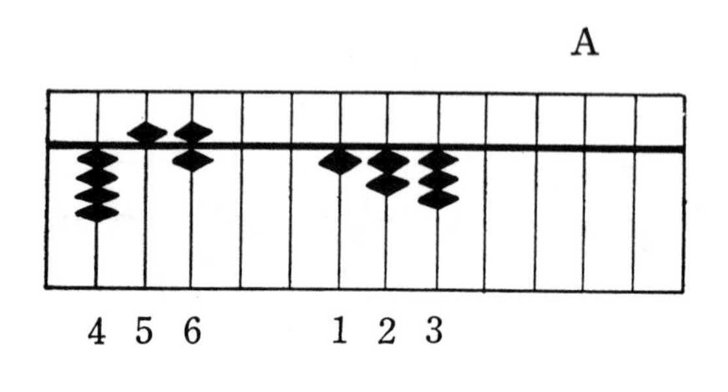

456	multiplier
123	multiplicand
56088	product

As there are three uprights in the multiplier, the third upright to the right of the multiplicand is the unit position of the product.

After indicating our first multiplication, "3×6 is 18", the unit position advances one upright to the left for the next multiplication, "3×5 is 15." Continuing in this way we complete the multiplication.

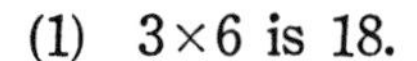

(1) 3×6 is 18.

①②3	multiplicand
④⑤6	multiplier
1 8	product

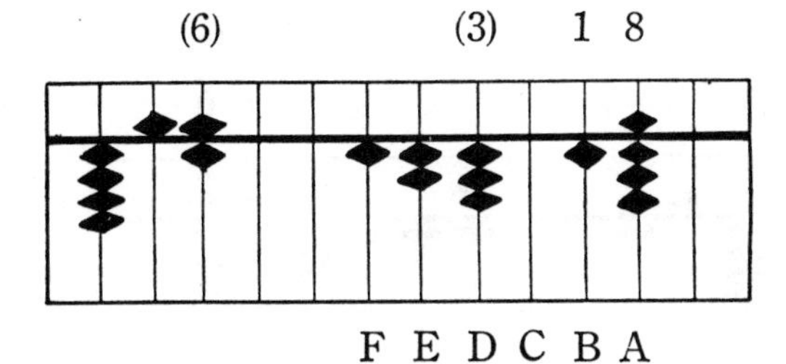

A upright is the unit position of the product of 3×6.

(2) 3×5 is 15.

①②3	multiplicand
④5⑥	multiplier
1 5 0	product

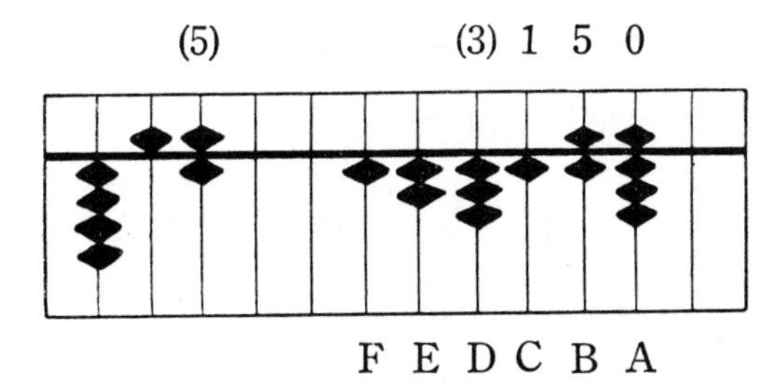

B upright is the unit position for the product of 3×5.

(3) 3×4 is 12.

①②3 multiplicand
4⑤⑥ multiplier
1 2 0 0 product

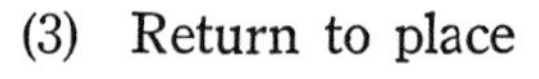

(3) Return to place

(4) 1 2 0 0

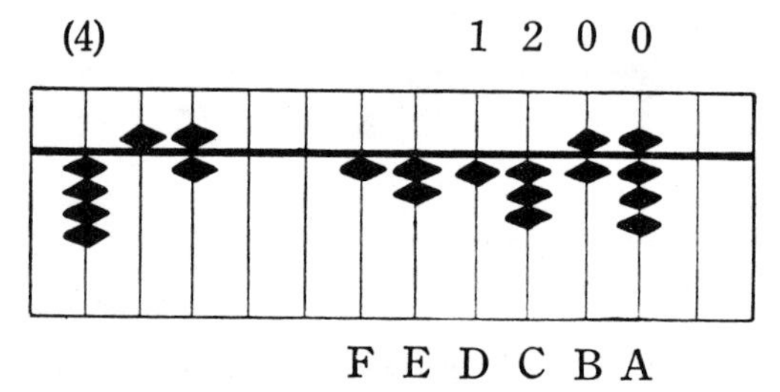

C upright is the unit position of the product of 3×4.

Thus we have added the (1) and (2) and (3)—18 and 150 and 1200, so 1368 should be indicated on the board.

The figure 3 which is the end figure of the multiplicand has now been multiplied by every figure of the multiplier so it must be returned to place.

(4) 2×6 is 12.

①2③ multiplicand
④⑤6 multiplier
1 2 0 product

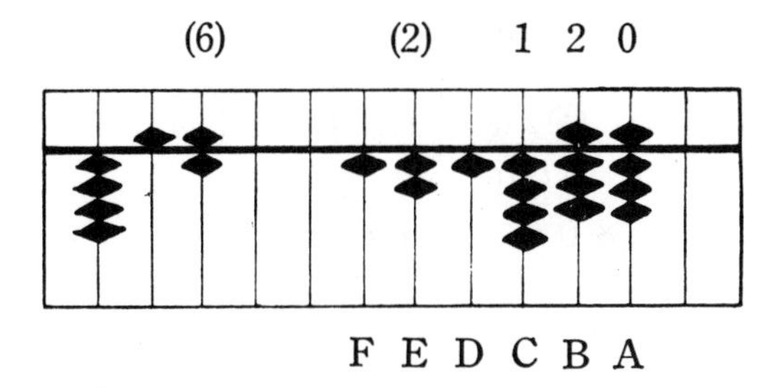

B is the upright for the unit position of the product of 2×6.

Adding the (1) and (2) and (3) and (4) 121488 is indicated on the board.

(5) 2×5 is 10.

①2③	multiplicand
④5⑥	multiplier
1 0 0 0	product

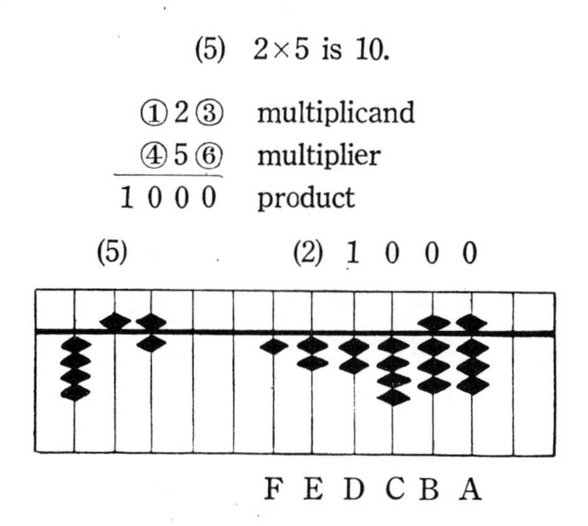

C upright is the unit position of the product of 2×5.

Adding the (1) and (2) and (3) and (4) and (5) we get 122488 indicated on the board.

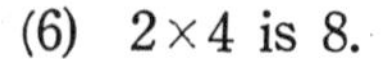

(6) 2×4 is 8.

①2③	multiplicand
4⑤⑥	multipiler
8 0 0 0	product

(2) Return to place.

(4) 8 0 0 0

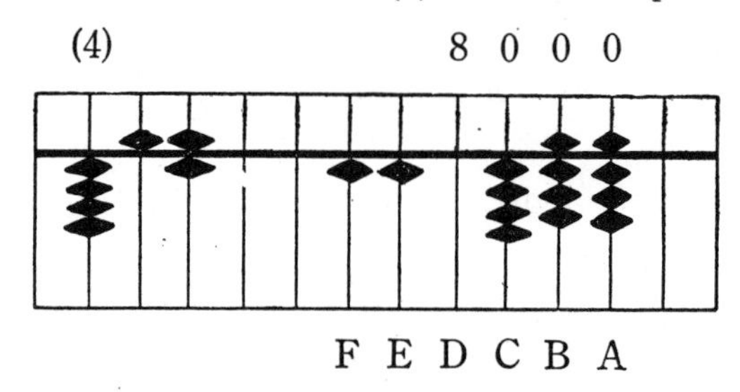

F E D C B A

D upright is the unit position of the product of 2×4.

Return to place the two Earth counters which indicate the figure 2 of the multiplicand as all multiplications with that figure have been completed.

There are two Earth counters already in position on the upright which is the unit position for our present multiplication; so instead of adding eight to the two think, "2 and 8 are 10". Return to place the two Earth counters and raise one Earth counter on the next upright to the left. Now having added the (1) and (2) and (3) and (4) and (5) and (6) we should have 110488 indicated on the board.

(7) 1×6 is 6.

1②③	multiplicand
④⑤6	multiplier
6 0 0	product.

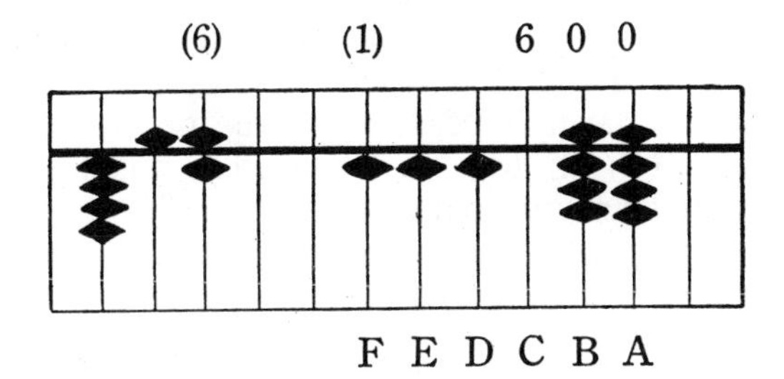

C is the unit upright for the product of 1×6.

But there are already 4 Earth counters in position on C upright; so thinking, "4 and 6 are 10", replace the four Earth counters and raise one Earth counter on the next upright to the left.

Adding the (1), (2), (3), (4), (5), (6), (7) we should have 111088 indicated on the board.

(8) 1×5 is 5.

1②③	multiplicand
④5⑥	multiplier
5 0 0 0	product

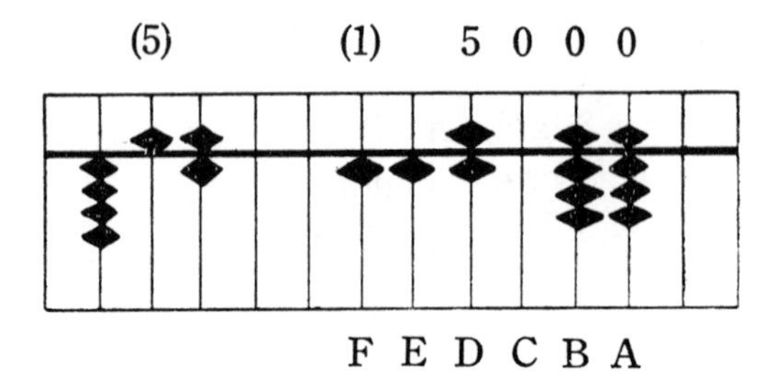

D upright is the unit position for the product of 1×5.

Adding the (1), (2), (3), (4), (5), (6), (7), and (8) we have 116088 indicated on the board.

(9) 1×4 is 4.

1 ②③	multiplicand
4 ⑤⑥	multiplier
4 0 0 0 0	product

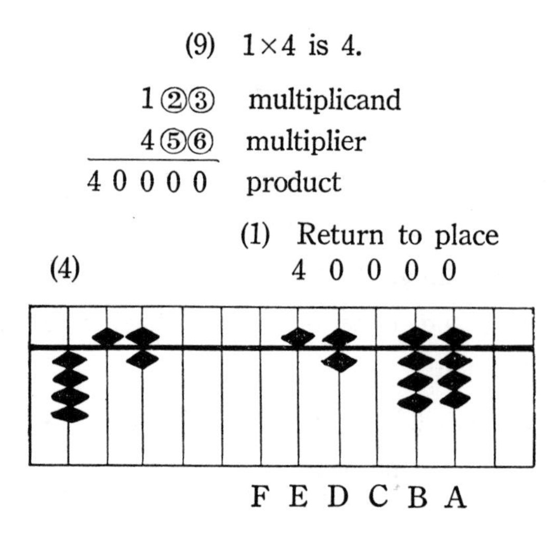

E upright is the unit position for the product of 1 and 4.

Prior to placing this product on the board we must replace the figure 1 of the Multiplicand as we have no further use for it.

Adding the (1), (2), (3), (4), (5), (6), (7), (8), and (9) we have 56088 on the board.

The process of multiplication just completed was as follows:

(1)	① ② 3		
	④ ⑤ 6	3 × 6	18
(2)	① ② 3		
	④ 5 ⑥	3 × 50	150
(3)	① ② 3		
	4 ⑤ ⑥	3 × 400	1200
(4)	① 2 ③		
	④ ⑤ 6	20 × 6	120
(5)	① 2 ③		
	④ 5 ⑥	20 × 50	1000
(6)	① 2 ③		
	4 ⑤ ⑥	20 × 400	8000
(7)	1 ② ③		
	④ ⑤ 6	100 × 6	600
(8)	1 ② ③		
	④ 5 ⑥	100 × 50	5000
(9)	1 ② ③		
	4 ⑤ ⑥	100 × 400	40000
			56088

A few more illustrations may help to make the process of multiplication clearer.

(1) When we multiply by a number of one place the next upright to the right of the multiplicand becomes the unit upright for the product, as is shown in the following illustration:

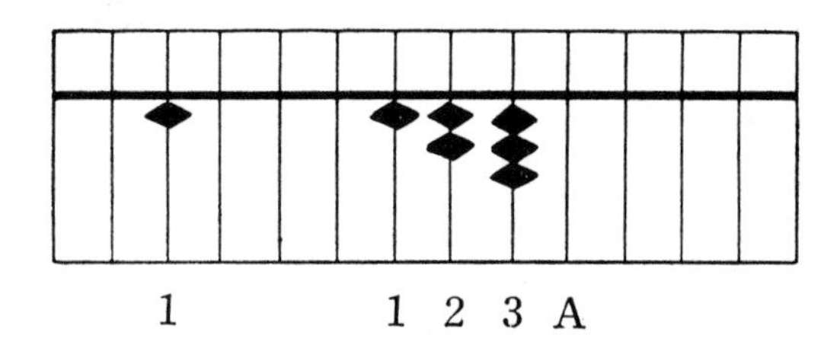

The upright A is the unit position of the product.

(2) When we multiply by a number of two places, the next upright to the right of the multiplicand becomes the position for 10's in the product, as is shown in the following illustration:

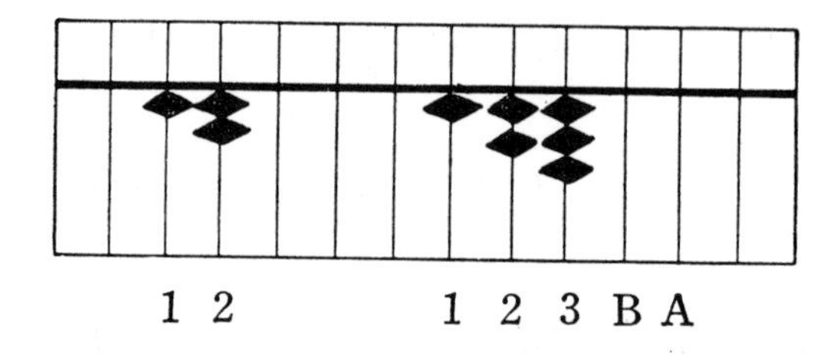

The upright B is the position for 10's in the product, and of course, A is the upright for the unit position.

(3) When we multiply by a number of three places, the next upright to the right of the multiplicand is the position for 100's in the product, as is shown in the following illustration:

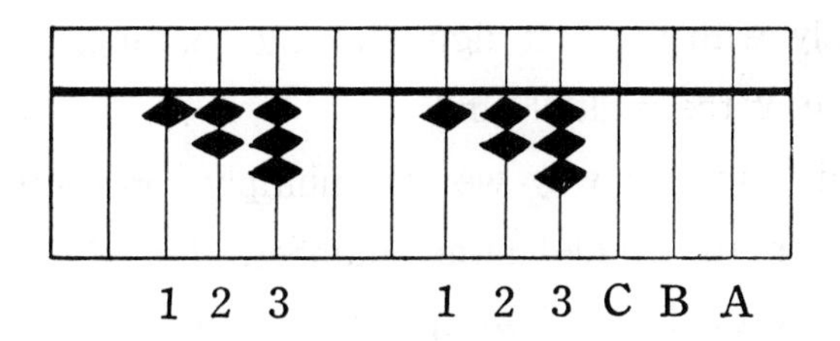

The upright C is the position for 100's in the product, and, of course, B is for 10's and A is for ones.

These illustrations should be sufficient to make this point clear.

Section 4.

Multiplication Beginning at the Left.

"Multiplication beginning at the Left" is contrary to the method we have just discussed. In this method we begin to multiply with the first figure of the multiplier and the last figure of the multiplicand.

Beginning in this way we can multiply consecutively, so it is easier to handle large numbers by Multiplication Beginning at the Left. Also, it is not so difficult to determine the unit position of the product.

Part 1.

The Process of Multiplication.

(1) Just as in the other method we must indicate the multiplier on the left hand side of the board, and indicate the multiplicand to the right of it.

(2) By the Multiplication Table we multiply the first figure of the multiplier by the last figure of the multiplicand.

(3) Before we indicate this product on the board, we must return to place the counters on the end upright of the multiplicand, which is the position for 10's in the product.

(4) Next we multiply the end figure of the multiplicand by the second figure of the multiplier and indicate this product on the second upright to the right of the multiplicand, continuing in this way till we reach the end figure of the multiplier.

Part 2.

The Determination of the Unit Position.

The determination of the Unit position in "Multiplication Beginning at the Left" is very much like the way of determining it in the previous method. Before beginning to multiply we decide that the unit position of the multiplicand is the position for 10's in the product.

If we do this and place the counters each time on the next righthand upright, the last one in the product becomes the unit position.

If we have zeros on the end uprights of both numbers we must count off as many uprights to the right as there are zeros.

I will show you by illustrations.

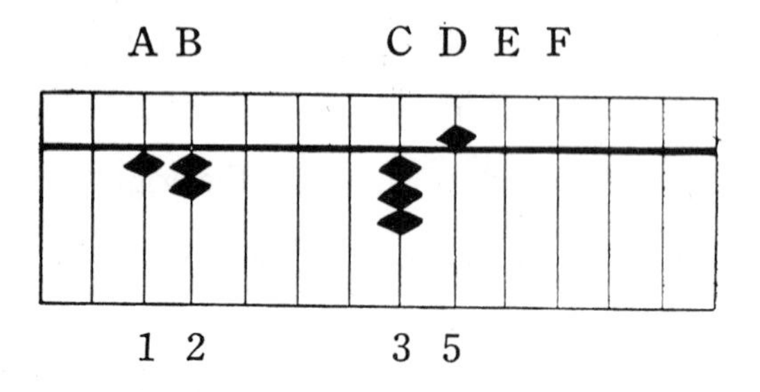

35 multiplicand

12 multiplier

When we multiply the one on the A upright (the first figure of the multiplier) by the 5 on the D upright (the end figure of the multiplicand), the D upright is the position for 10's and the E upright is the position for ones.

When we multiply the 2 on the B upright by the 5 on the D upright, the E upright is the position for tens in the product, and the F upright is the unit position.

When we multiply the 1 on the A upright by the 3 on the C upright, the C upright is the position for 10's, and the D upright is the unit position.

When we multiply the 2 on the B upright by the 3 on the C upright, the D upright is the position for 10's and the E is the unit position.

Part 3.

Example 1. 56 multiplicand
12 multiplier
672 product

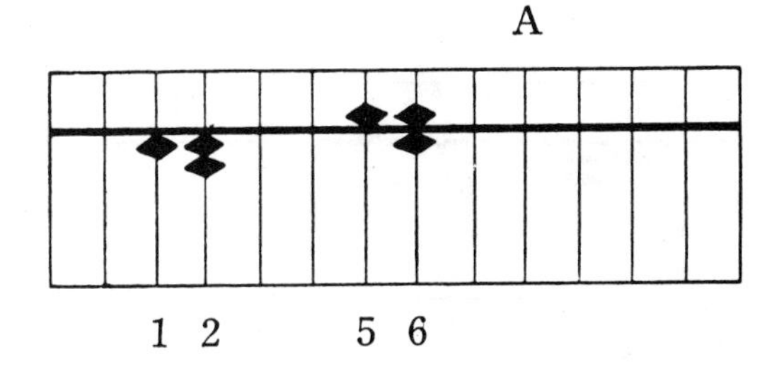

A is the unit position of the product.

(1) Place the multiplier on the lefthand side of the board and the multiplicand on the right of it, as is shown in the above illustration.

(2) As there are two uprights in the multiplier, count off two places from the multiplicand and you will find the unit position.

(3) Multiply the 6, which is the end figure of the multiplicand, by the 1, which is the first figure of the multiplier. This gives us, "1×6 is 6."

(4) Before indicating this product we must return to place the figure 6 of the multiplicand and then press between the

fingers the Heaven counter and one Earth counter on the next upright to the right. 5060 should be indicated on the board.

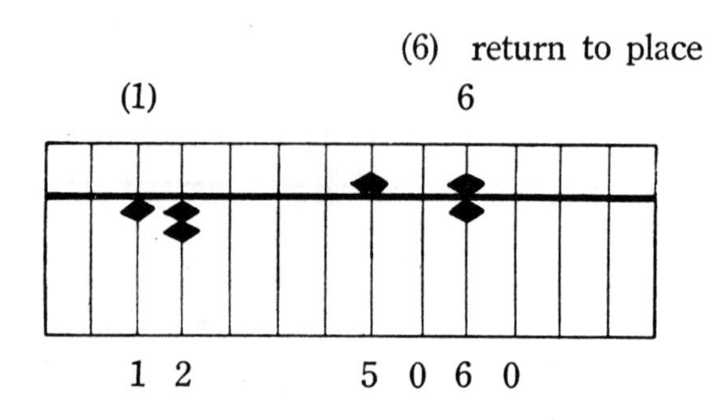

(5) Although we have returned to place the end figure of the multiplicand we must not forget it, as we have not yet completed our multiplications.

(6) We must now multiply the figure 6, which we returned to place, and the second figure of the multiplier. This gives us, "6×2 is 12."

(7) Raise one Earth counter on the next upright to the right, which is the position for 10's in the product, and two Earth counters on the unit upright. 5072 should now be indicated on the board.

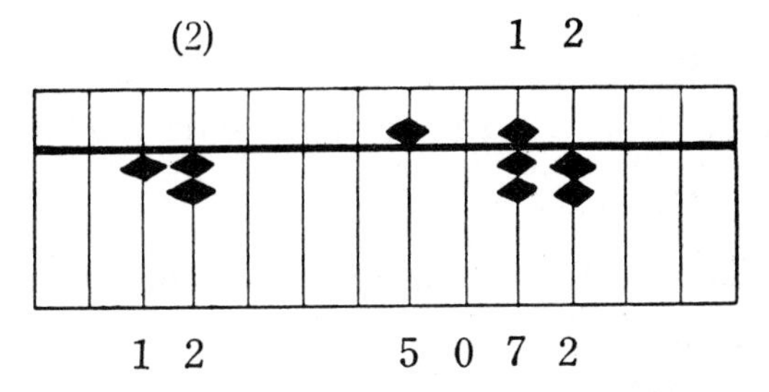

(8) We have now completed our multiplications with the end figure of the multiplicand, having multiplied it by every figure in the multiplier, beginning with the lefthand figure. We must proceed in this way with each figure of the multiplicand.

(9) Multiplying the second figure of the multiplicand, 5, with the first figure of the multiplier, 1, we get, "5×1 is 5".

(10) Before placing this product we must return to place the figure 5 of the multiplicand. Slide down the 5 counter of the next upright to the right, the position for 100's in the product. 572 should be indicated on the board.

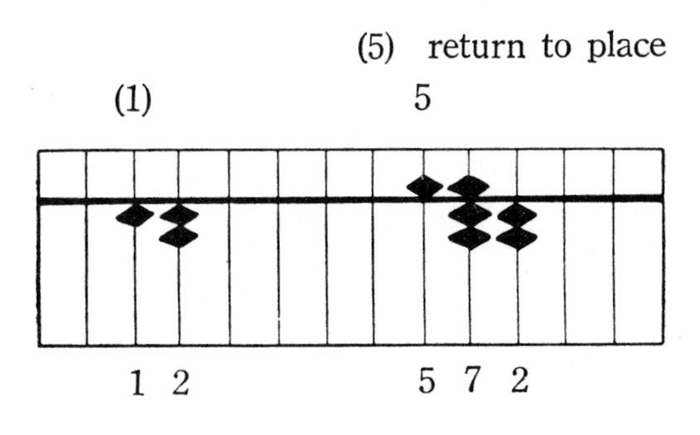

(11) We must keep in mind that the figure which we returned to place in the multiplicand was 5, as we must now multiply it by the second figure of the multiplier. This gives us, "5×2 is 10."

(12) Raise cne Earth counter on the upright for 100's in the product.

672 is indicated on the board. It is the product.

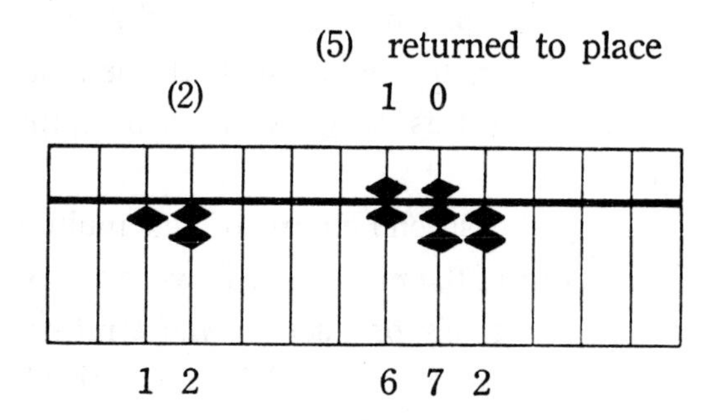

Example 2. 142 multiplicand
95 multiplier
13490 product

Example 2.	142	multiplicand
	95	multiplier
	13490	product

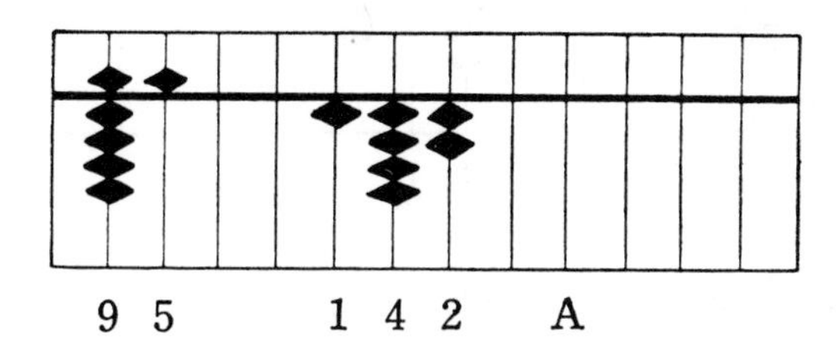

(1) Indicate the multiplier on the lefthand side of the board and the multiplicand to the right of it, as shown in the above illustration.

(2) As there are two uprights in the multiplier, count off two places to the right of the multiplicand to determine the unit position for the product.

(3) Multiplying the end figure of the multiplicand by the first figure of the multiplier, we get, "2×9 are 18."

(4) Return to place the Earth counters which indicate the figure 2 of the multiplicand. Raise one Earth counter on the same upright to indicate the 10 of the product 18, and on the next upright to the right press between the fingers the 5 counter and three 1 counters, to indicate the 8.

1418 is now indicated on the board.

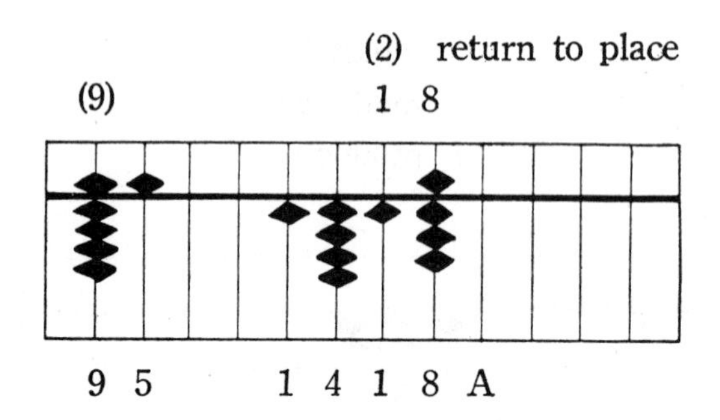

(5) We must now multiply the figure 2, which we have returned to place, and the second figure of the multiplier. This gives us, "2×5 is 10."

(6) Raise one Earth counter on the upright for 10's in the product. 1419 should now be indicated on the board.

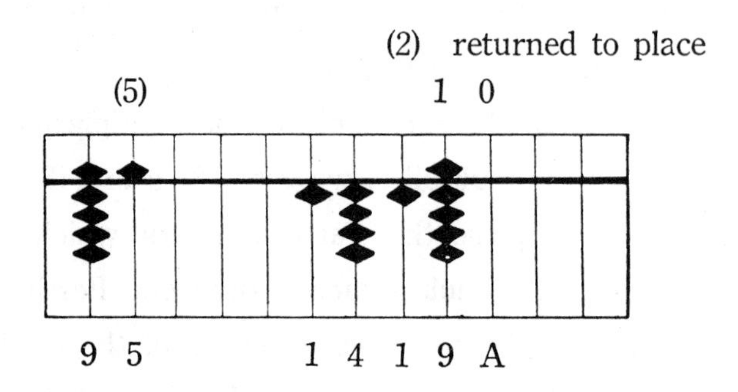

(7) Multiplying the figure 4 of the multiplicand by the first figure of the multiplier, we get, "4×9 is 36."

(8) Before placing this product 36, we must return to place the counters which indicate the figure 4 of the multiplicand. Then raise three Earth counters on the same upright and indicate 6 on the next upright to the right. The board should now indicate 13790.

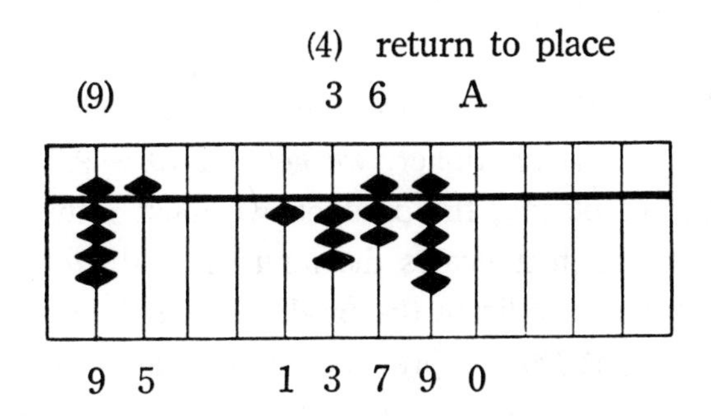

(9) Multiplying the figure 4, which we have returned to place, by the second figure of the multiplier, we get, "4×5 is 20."

(10) Raise two Earth counters on the upright for 100's in the product. 13990 should now be indicated on the board.

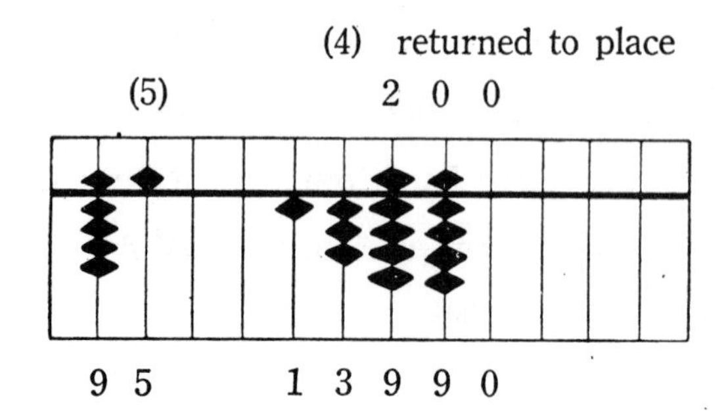

(11) Multiplying the figure 1 of the multipllcand by the first figure of the multiplier, we get, " 1×9 is 9."

(12) Before placing the product, we must return to place the counter which indicates the figure 1 of the multiplicand. We now wish to indicate the product 9 on the next upright to the right, but from a previous multiplication 3 is already indicated there. Thinking, A. T " 9 and 1 are 10 ", return to place one of these three counters and at the same time raise one Earth counter on the next upright to the left. 12990 is now indicated on the board.

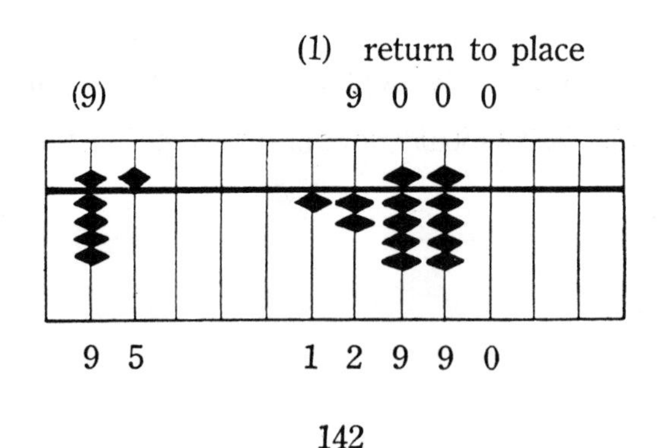

(13) Keeping in mind the figure 1 of the multiplicand, which we have returned to place, we multiply it by the second figure ot the multiplier. This gives us, "1×5 is 5."

(14) We now wish to indicate this product on the upright for 100's in the product, but 9 is already indicated there. Therefore, we think, "5 and 5 is 10". And return to place the Heaven counter on that upright and raise one Earth counter on the next upright to the left.

13490 should now be indicated on the board and as we have completed all the multiplications, it is the final product.

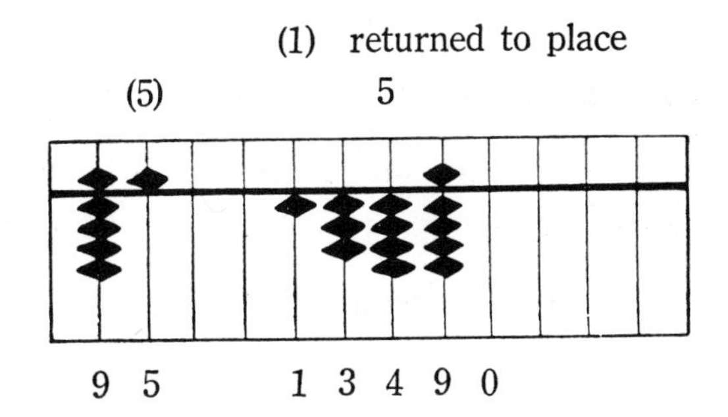

Example 3.

2854	multiplicand
102	multiplier
291108	product

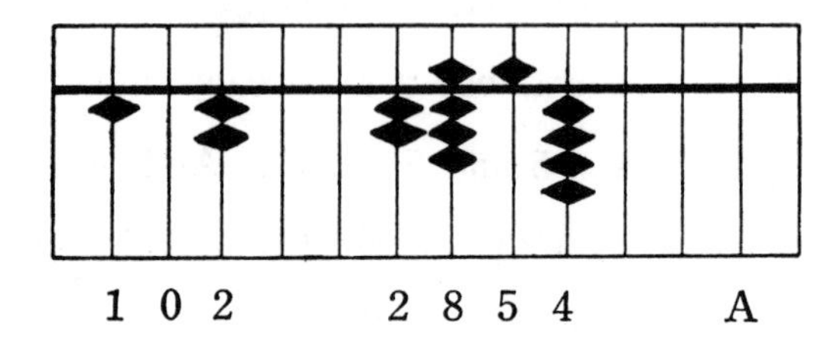

(1) Indicate the multiplier on the lefthand side of the board and the multiplicand to the right of it as is shown in the above illustration.

(2) As there are three figures in the multiplier, the third upright to the right of the multiplicand is the unit upright for the product.

(3) Multiplying the end figure of the multiplicand by the first figure of the multiplier, we get, "1×4 is 4." Before indicating this product we must return to place the counters which indicate the end figure of the multiplicand and then place our product 4 on the next upright to the right. The board should now indicate 28504.

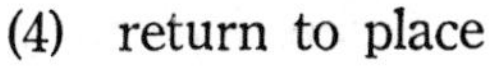

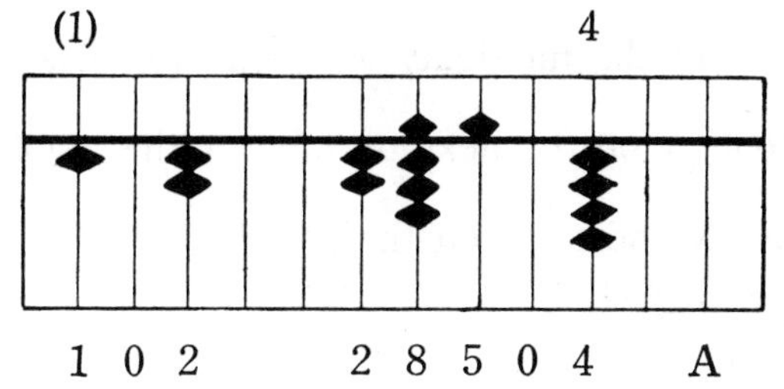

(5) Keeping in mind the figure 4 which we have returned to place we multiply it by the end figure of the multiplier. This gives us, "4×2 is 8."

(6) Press between the fingers the Heaven counter and three Earth counters on the unit position of the product. 2850408 should now be indicated on the board.

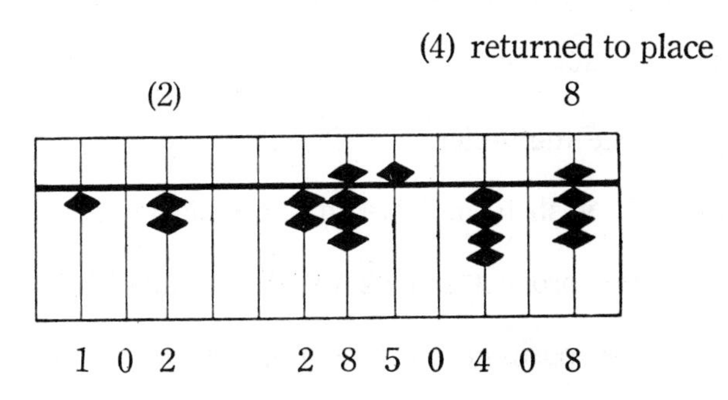

(7) Multiplying the second figure of the multiplicand with the first figure of the multiplier we get "5×1 is 5."

(8) Before indicating this product we must return to place the figure 5 of the multiplicand, then we slide down the Heaven counter on the next upright to the right. 2805408 should now be indicated on the board.

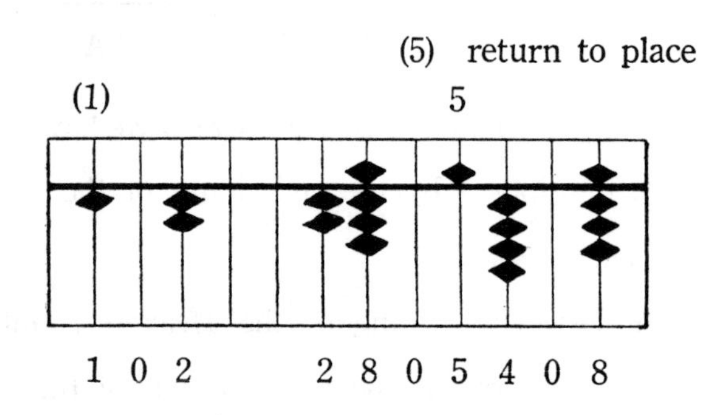

(9) Keeping in mind the figure 5 of the multiplicand, which we have returned to place, we multiply it by the end figure of the multiplier, "5×2 is 10."

(10) We wish to indicate this product on the upright for 100's in the product but as there are already four Earth counters in position on that upright we slide down the Heaven counter and return the four Earth counters. 2805508 should now be indicated on the board.

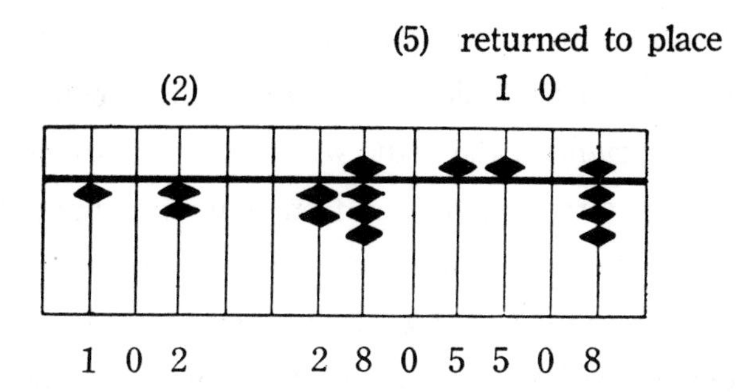

(11) Multiplying the third figure of the multiplicand by the first figure of the multiplier we get, "8×1 is 8."

(12) Before indicating this product we return to place the counters representing the figure 8 of the multiplicand, then we indicate the product 8 on the next upright to the right. 2085508 should now be indicated on the board.

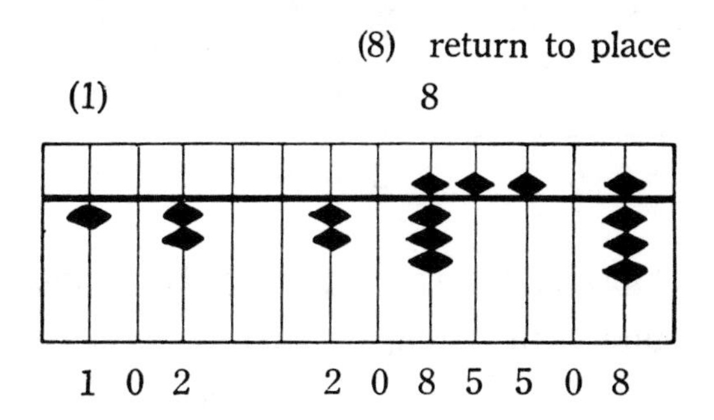

(13) Keeping in mind the figure 8 which we have returned to place multiply it by the last figure of the multiplier. This gives us, "8×2 is 16."

(14) Raise one Earth counter on the upright for 1000's in the prcduct. This indicates the 10 of our product 16. On the next upright to the right where we wish to indicate the 6 there is 5 already indicated from a previous multiplication. Therefore we must think, " 5 and 5 are 10 ", and raising one Earth counter on the 100 upright, return the Heaven counter on the 100 upright to place and raise one Earth counter on the next upright to the left. 2087108 should now be indicated on the board.

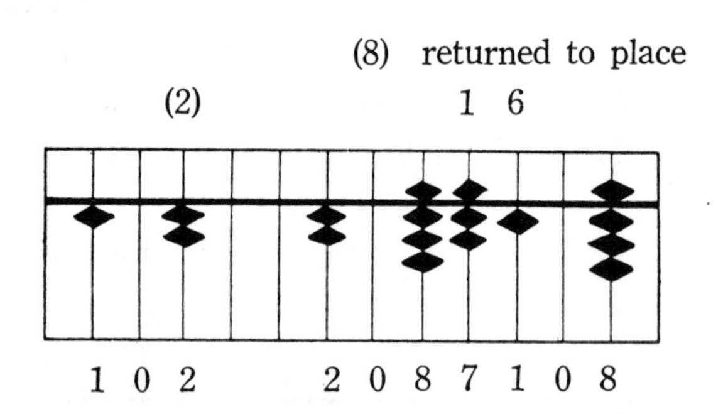

(15) Multiplying the first figure of the multiplicand by the first figure of the multiplier we get, " 2×1 is 2."

(16) Return to place the counters which are used to indicate the figure 2 of the multiplicand, and raise two Earth counters on the next upright to the right. 287108 should be indicated on the board.

(17) Keeping in mind the figure 2 of the multiplicand we multiply it by the last figure of the multiplier. This gives us, "2×2 is 4."

(18) We wish to indicate this product on the upright for 1000's in the product but as it already indicates 7 we think, "4 and 6 are 10" and return to place the Heaven counter and one Earth counter on the upright, at the same time raising one Earth counter on the next upright to the left.

291108 is indicated and it is the final product.

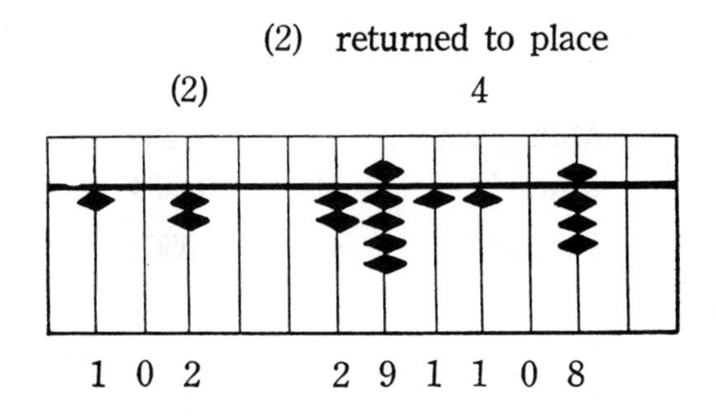

Cover the product in row A. Multiply and write the products. Compare your results with the printed products.

Do again any which were wrong. Do the same with row B and the other rows.

Row A.

Multiplicand	316	423	265
Multiplier	24	19	27
Product	7584	8037	7155

Row B.

Multiplicand	95	127	328
Multiplier	76	56	234
Product	7220	7112	76752

Row C.

Multiplicand	29	103	235
Multiplier	14	54	25
Product	406	5562	5875

Row D.

Multiplicand	$ 7.24	$ 8.06
Multiplier	86	92
Product	$ 622.64	$ 741.52

Row E.

Multiplicand	$ 64.76	$ 76.32
Multiplier	84	89
Product	$ 5439.84	$ 6792.48

Section 5.

Multiplying with Decimals.

There is no difference in the process of multiplication for decimals and ordinary numbers. The only difference is in determining the unit position.

(1) When the multiplier has no zero directly below the decimal point, we consider the unit position of the multiplicand, the unit position for the product.

(2) When the multiplier has two zeros below the decimal point, we count off two places from the unit position of the multiplicand to the left. This is the unit position of the product.

Some illustrations will make this point clear.

Example 1.

2345 multiplicand

0.123 multiplier

The upright A is the unit position of the product.

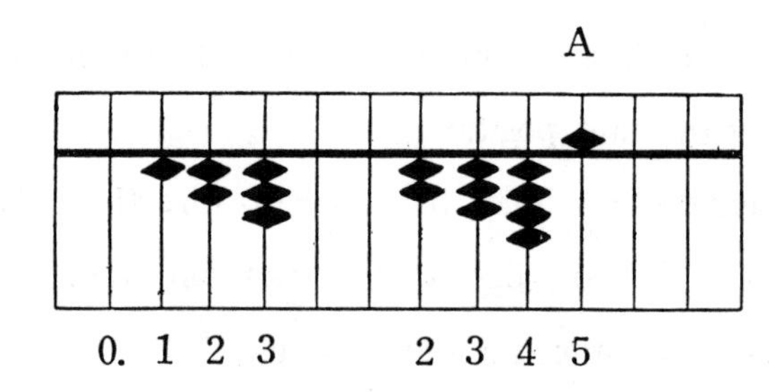

Example 2.

2345 multiplicand

0.00123 multiplier

The upright A is the unit position for the product.

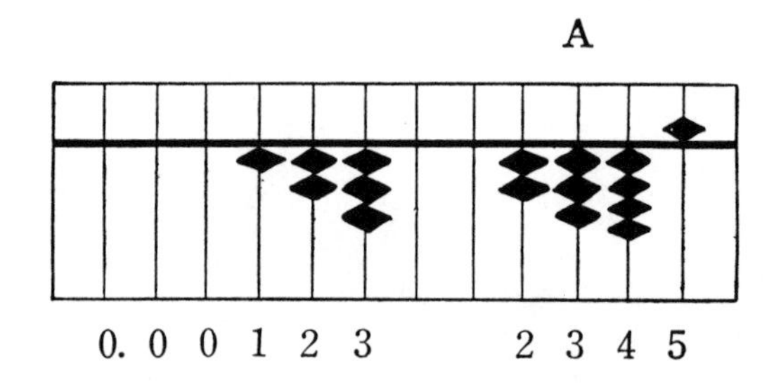

Section 6.

Multiplying Tenths and Hundredths.

You can multiply with decimals, which includes numbers that represent dollars and cents, just as you multiply integers, if you put the decimal point where it belongs in the product.

$$23.45 \times 52 =$$

After you have multiplied with the decimals just as you would with whole numbers, point off as many decimal places in the product as there are in the multiplier and the multiplicand together.

Find the product.

Row A.	Row B.
(1) $138 \times 2 =$	(1) $682 \times 6 =$
(2) $243 \times 3 =$	(2) $737 \times 7 =$
(3) $358 \times 4 =$	(3) $685 \times 8 =$
(4) $519 \times 5 =$	(4) $946 \times 9 =$
(5) $472 \times 6 =$	(5) $478 \times 2 =$

Cover the answer with a card. State each product. Then look at the answer to make sure that you were right.

(1) $604 \times 39 = 23556$

(2) $746 \times 41 = 30586$

(3) $824 \times 48 = 39552$

(4) $983 \times 52 = 51116$

(5) $198 \times 57 = 11286$

(6) $217 \times 64 = 13888$

(7) $362 \times 65 = 23530$

(8) $439 \times 73 = 32047$

(9) $573 \times 79 = 45267$

(10) $661 \times 81 = 53541$

(11) $724 \times 86 = 62264$

(12) $806 \times 92 = 74152$

(13) $955 \times 98 = 93590$

(14) $1647 \times 13 = 21411$

(15) $2838 \times 16 = 45408$

(16) $\$2.63 \times 47 = \$123.61$

(17) $\$5.03 \times 42 = \$211.26$

(18) $\$9.25 \times 38 = \$351.50$

(19) $\$6.74 \times 81 = \$545.94$

(20) $\$3.27 \times 29 = \$94.83$

(21) $\$5.54 \times 48 = \$265.92$

(22) $\$1.32 \times 56 = \$73.92$

(23) $\$2.25 \times 47 = \$105.75$

(24) $\$4.55 \times 32 = \$145.60$

(25) $\$4.01 \times 58 = \$232.58$

(26) $\$5.83 \times 49 = \$285.67$

(27) $\$7.11 \times 31 = \$220.41$

(28) $\$8.25 \times 53 = \$437.25$

(29) $\$9.03 \times 12 = \$108.36$

(30) $\$3.42 \times 48 = \$164.16$

CHAPTER V.

DIVISION

There are two methods of division in general use, " Division by Unification ", and " Division by Abbreviation."

Division by Unification.

In this method both Division and Multiplication Tables are used.

Division by Abbreviation.

In this method only Multiplication Table is used in solving the problems.

$$\text{Divisor} \quad 28\overline{)1232} \quad \text{dividend} \qquad (44 \text{ quotient})$$

Section 1.

Division by Unification.

This is the method in which we use both Multiplication and Division Tables when we divide the dividend by the divisor.

By this method we can find out on the abacus how many times a certain number is contained in another number.

If you memorize the Division Table you will find this new method of division more simple than division in writing.

Part 1.

Division Table by Unification.

Dividing by 1.

1	into	1	advance	1		1	into	6	advance	6
1	"	2	"	2		1	"	7	"	7
1	"	3	"	3		1	"	8	"	8
1	"	4	"	4		1	"	9	"	9
1	"	5	"	5						

Illustration of Division by 1.

6 quotient

Divisor 1) 6 dividend

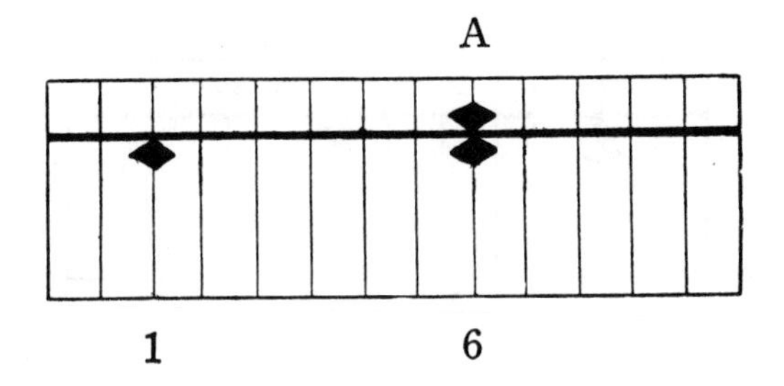

Dividing the 6 on the A upright by the 1, we get, "1 into 6 advance 6." We return to place the counters which indicate 6 and on the next upright to the left indicate the quotient 6.

The entire Division Table of 1 is done in this way, advancing one upright to the left for the quotient.

Dividing by 2.

2	into	1	Heaven	5
2	into	2	advance	1

Illustrations of Division by 2.

.5 quotient
Divisor 2) 1 dividend

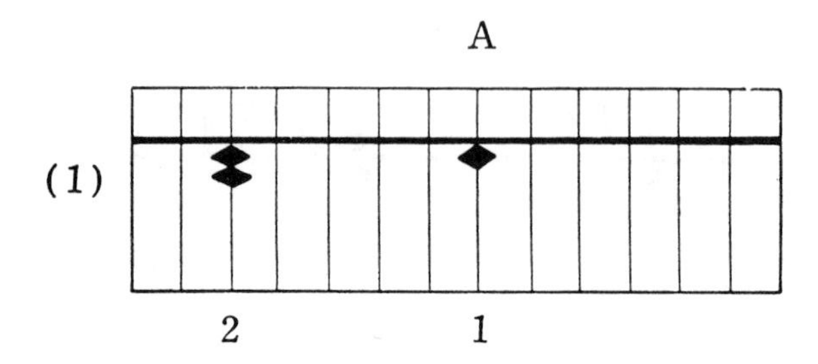

When we divide the 1 on the A upright by 2, we get, "2 into 1 Heaven 5." Return to place the Earth counter on the A upright and slide down the Heaven counter on the same upright. This indicates .5 on the board.

1 quotient
Divisor 2) 2 dividend

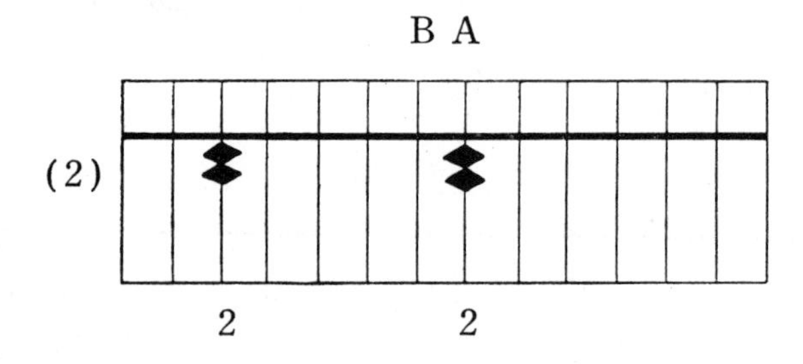

Dividing the 2 on the A upright by 2, we get, "2 into 2 advance 1."

Return to place the two Earth counters on the A upright and at the same time raise one Earth counter on the next upright to the left.

The quotient 1 is now indicated on the board.

Dividing by 3.

3 into 1	3 and 1 over
3 into 2	6 and 2 over
3 into 3	advance 1

Illustrations of Division by 3.

3 into 1 3 and 1 over

3 and 1 over quotient

Divisor 3) 1 dividend

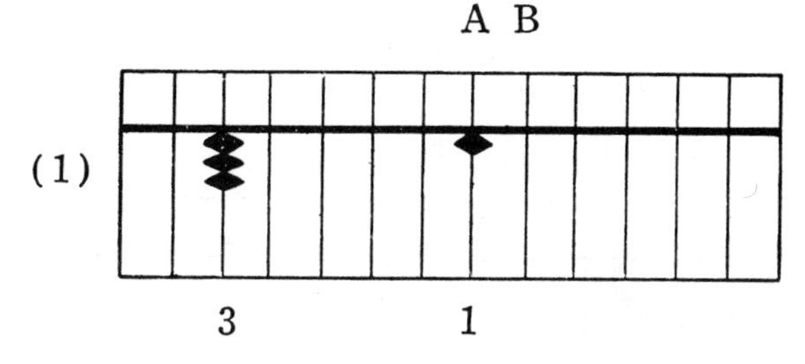

When we divide the 1 on the A upright by 3, we get, "3 into 1, 3 and 1 over." Instead of returning to place the Earth counter on the A upright, add two counters to it and raise one Earth counter on the next upright to the right, the B upright.

31 is indicated on the board.

$$\begin{array}{r l} & \underline{\ 6\ } \text{ and 2 over quotient} \\ \text{Divisor } 3) & 2 \text{ dividend} \end{array}$$

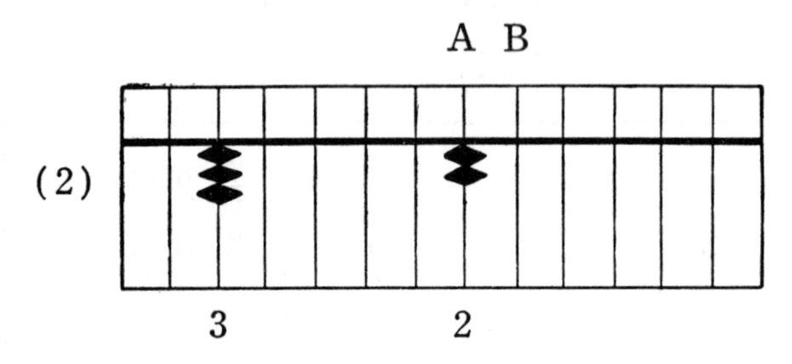

When we divide the 2 on the A upright by 3, we get, "3 into 2, 6 and 2 over." Instead of returning to place the two Earth counters on the A upright, add four Earth counters to them and raise two Earth counters on the B upright, which is the next upright to the right.

62 is indicated on the board.

$$\begin{array}{r l} & \underline{\ 1\ } \text{ quotient} \\ \text{Divisor } 3) & 3 \text{ dividend} \end{array}$$

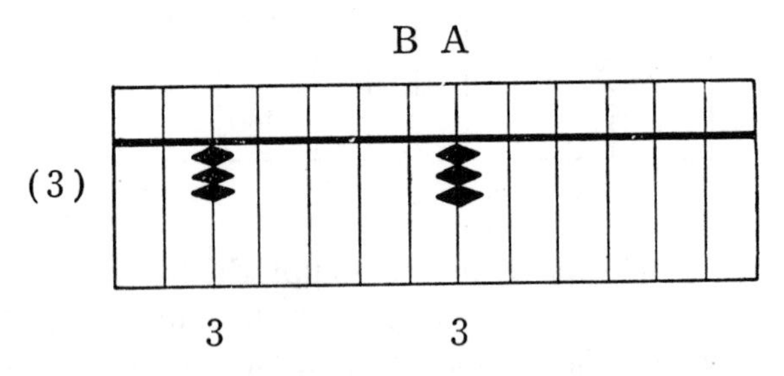

When we divide the 3 on the A upright by 3, we get, "3 into 3 advance 1."

Return to place the three Earth counters on the A upright, and at the same time raise one Earth counter on the next upright to the left, or the B upright.

1 is indicated on the board.

Dividing by 4.

4 into 1	2 and 2 over
4 into 2	Heaven 5
4 into 3	7 and 2 over
4 into 4	advance 1

Illustrations of Division by 4.

$$\text{Divisor } 4\overline{)\,1\,}\text{ dividend} \quad \text{quotient: } 2 \text{ and } 2 \text{ over}$$

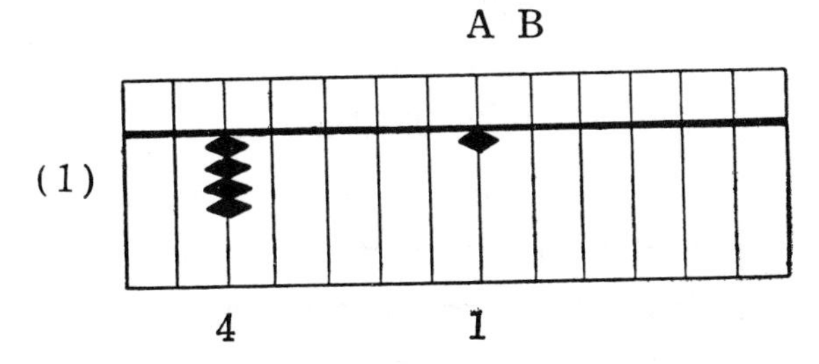

When we divide the 1 on the A upright by 4, we get, "4 into 1, 2 and 2 over." Instead of returning to place the 1 Earth counter on the A upright, add one Earth counter to it, then raise two Earth counters on the B upright, which is the next upright to the right.

22 is indicated on the board.

5 quotient
Divisor 4) 2 dividend

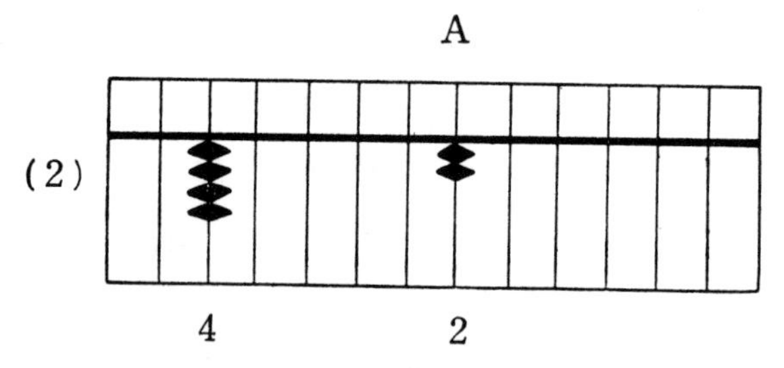

When we divide the 2 on the A upright by 4, we get, "4 into 2 Heaven 5." Return to place the Earth counters which are on the A upright and then slide down the Heaven counter, which is on the same upright.

5 is indicated on the board.

7 and 2 over quotient
Divisor 4) 3 dividend

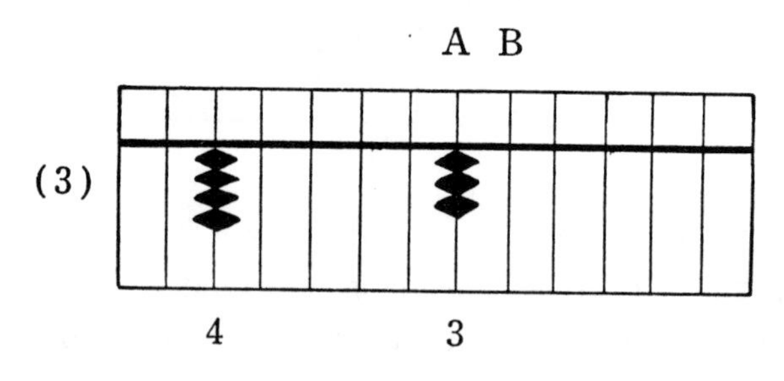

When we divide the 3 on the A upright by 4, we get, "4 into 3, 7 and 2 over." Instead of returning to place the 3 Earth counters on the A upright, add 4 Earth counters to this upright and raise two Earth counters on the next upright to the right, or the B upright.

72 is indicated on the board.

1 quotient
Divisor 4) 4 dividend

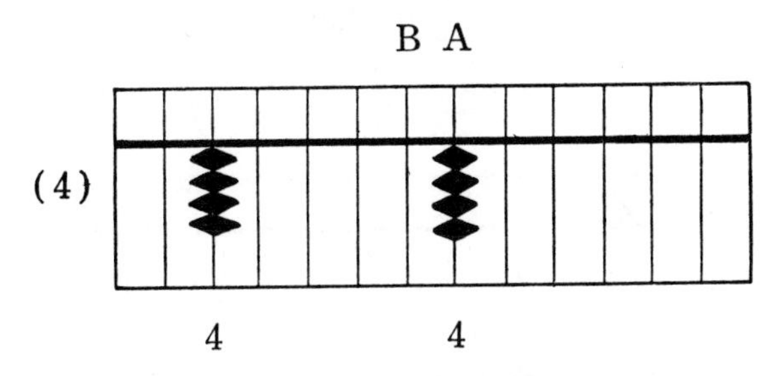

When we divide the 4 on the A upright by 4, we get, "4 into 4, advance 1." Return to place the four Earth counters on the A upright and raise one Earth counter on the next upright to the left, or the B upright.

1 is indicated on the board.

Dividing by 5.

5 into 1	plus 1
5 into 2	plus 2
5 into 3	plus 3
5 into 4	plus 4
5 into 5	advance 1

Illustrations of Division by 5.

$$\text{Divisor}\quad 5\overline{)\,1\,}^{\,2}\quad \begin{matrix}\text{quotient}\\ \text{dividend}\end{matrix}$$

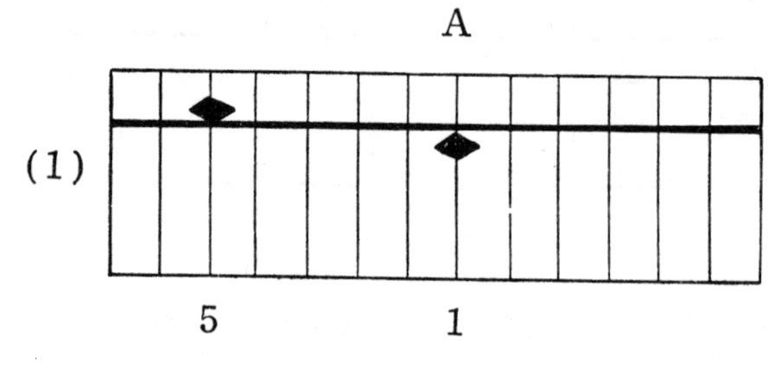

When we divide the 1 on the A upright by 5, we get, "5 into 1, plus 1." Add the 1 to the 1 Earth counter on the A upright.

2 is indicated on the board.

$$\text{Divisor}\quad 5\overline{)\,2\,}^{\,4}\quad \begin{matrix}\text{quotient}\\ \text{dividend}\end{matrix}$$

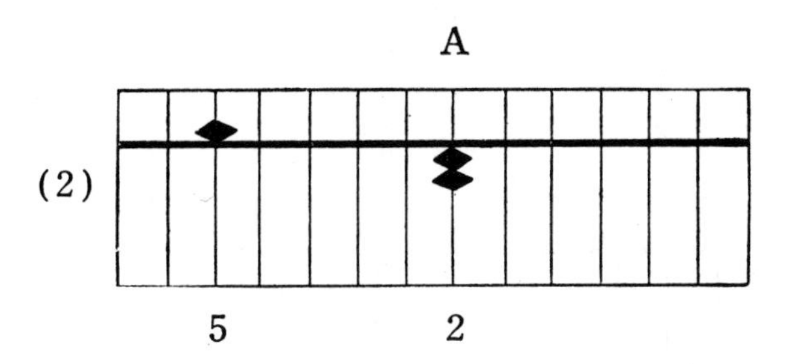

When we divide the 2 on the A upright by 5, we get, "5 into 2, plus 2." Add the 2 to the two Earth counters on the A upright.

4 is indicated on the board.

$$\text{Divisor}\quad 5\overline{)\,3\,}\ \text{dividend} \qquad \text{quotient } 6$$

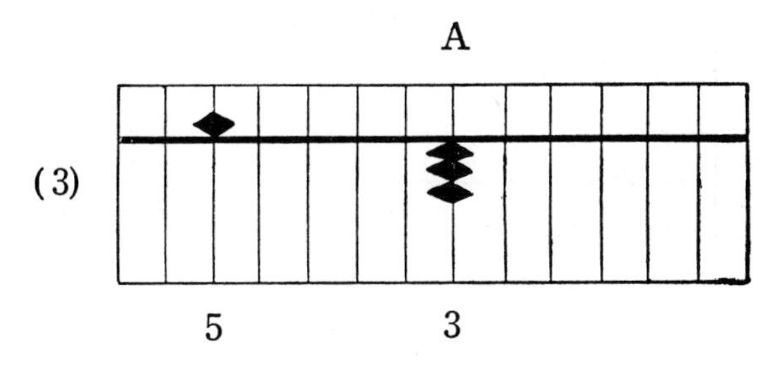

When we divide the 3 on the A upright by 5, we get, "5 into 3, plus 3." Add the 3 Earth counters on the A upright where 3 is already indicated.

6 is indicated on the board.

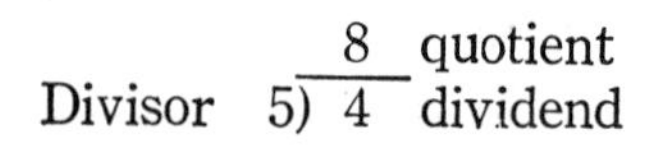

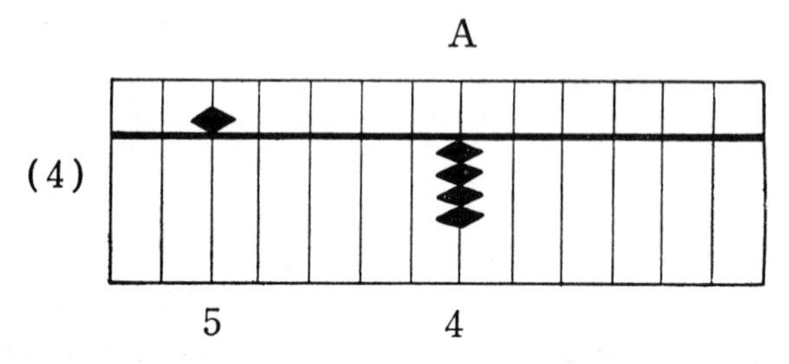

When we divide the 4 on the A upright by 5, we get, "5 into 4, plus 4." Add 4 Earth counters to the 4 already indicated on the A upright.

8 is indicated on the board.

1 quotient
Divisor 5) 5 dividend

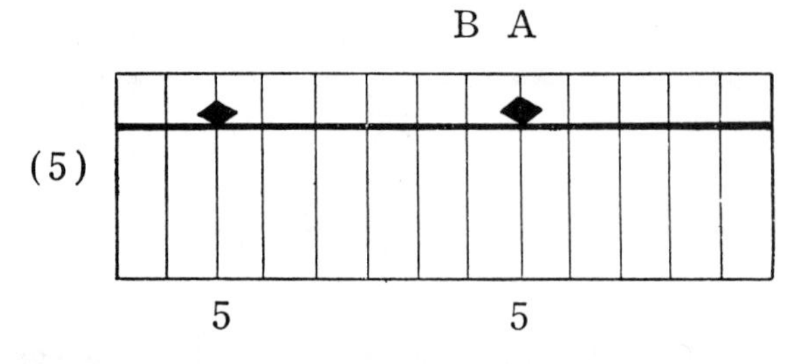

When we divide the 5 on the A upright by 5, we get, "5 into 5, advance 1." Return to place the Heaven counter on the A upright and raise one Earth counter on the next upright to the left, or the B upright.

1 is indicated on the board.

Dividing by 6.

6 into 1	1 and 4 over
6 into 2	3 and 2 over
6 into 3	Heaven 5
6 into 4	6 and 4 over
6 into 5	8 and 2 over
6 into 6	advance 1

Illustrations of Division by 6.

1 and 4 over quotient
Divisor 6) 1 dividend

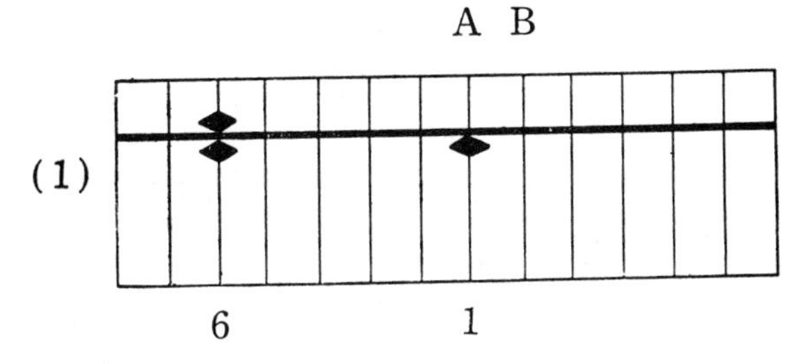

When we divide the 1 on the A upright by 6, we get, "6 into 1, 1 and 4 over." Raise 4 Earth counters on the next upright to the right, or the B upright.

14 is indicated on the board.

3 and 2 over quotient
Divisor 6) 2 dividend

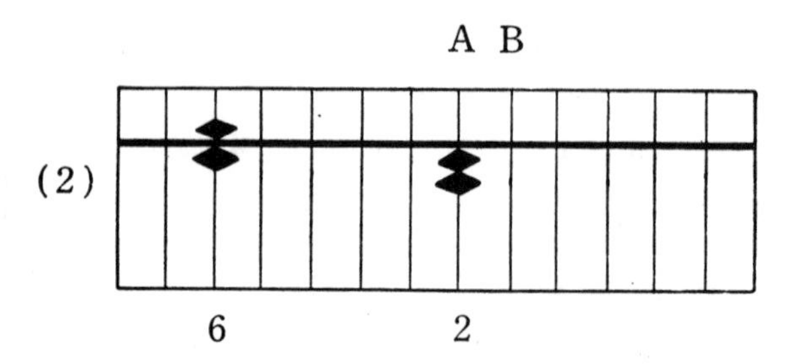

When we divide the 2 on the A upright by 6, we get, "6 into 2, 3 and 2 over." Instead of returning to place the two Earth counters on the A upright, add one Earth counter to them and raise two Earth counters on the next upright to the right, or the B upright.

32 is indicated on the board.

$$\text{Divisor} \quad 6\overline{)\,3\,}\; \begin{array}{l}\text{5 quotient}\\ \text{dividend}\end{array}$$

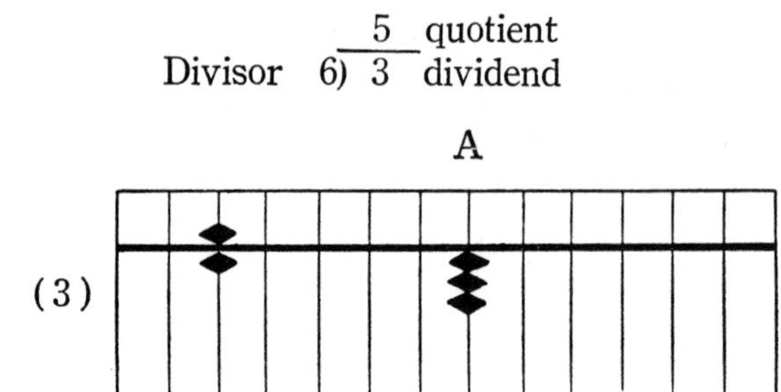

When we divide the 3 on the A upright by 6, we get, "6 into 3 Heaven 5." Return to place the three Earth counters on the A upright, and at the same time, slide down the Heaven counter on the same upright.

5 is indicated on the board.

6 and 4 over quotient
Divisor 6) 4 dividend

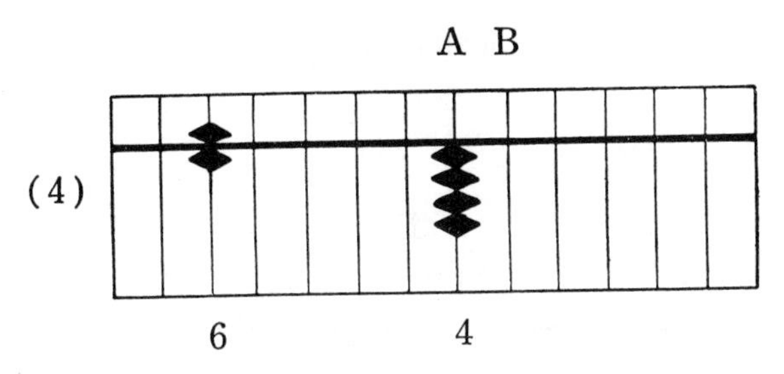

When we divide 4 by 6, we get, "6 into 4, 6 and 4 over." Instead of returning to place the four counters which are indicated on the A upright, add two Earth counters to this upright. Then raise four Earth counters on the next upriht to the right, the B upright.

64 is indicated on the board.

8 and 2 over quotient
Divisor 6) 5 dividend

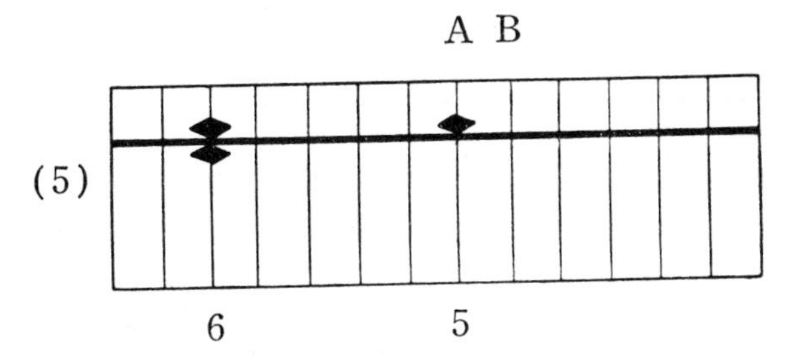

When we divide 5 on the A upright by 6, we get, "6 into 5, 8 and 2 over." Instead of returning to place the

Heaven counter on the A upright, add three Earth counters to it and then raise two Earth counters on the next upright to the right.

82 is indicated on the board.

             1   quotient
Divisor 6) 6   dividend

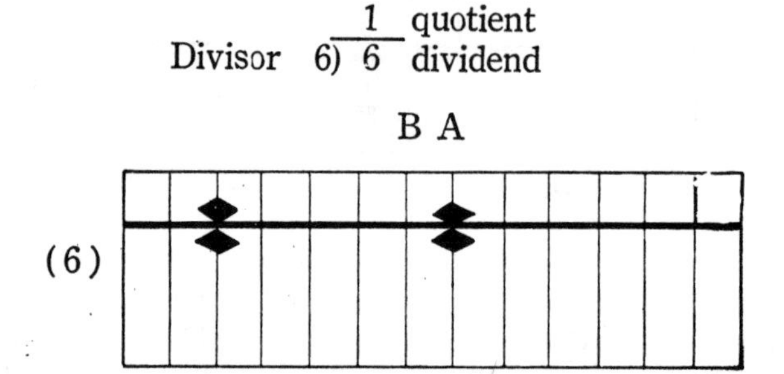

6 6

When we divide the 6 on the A upright by 6, we get, "6 into 6 advance 1." We return to place the counters which indicate 6 on the A upright and raise one Earth counter on the next upright to the left.

1 is indicated on the board.

Dividing by 7.

7 into 1	1 and 3 over
7 into 2	2 and 6 over
7 into 3	4 and 2 over
7 into 4	5 and 5 over
7 into 5	7 and 1 over
7 into 6	8 and 4 over
7 into 7	advance 1

Illustrations of Division by 7.

1 and 3 over quotient
Divisor 7) 1 dividend

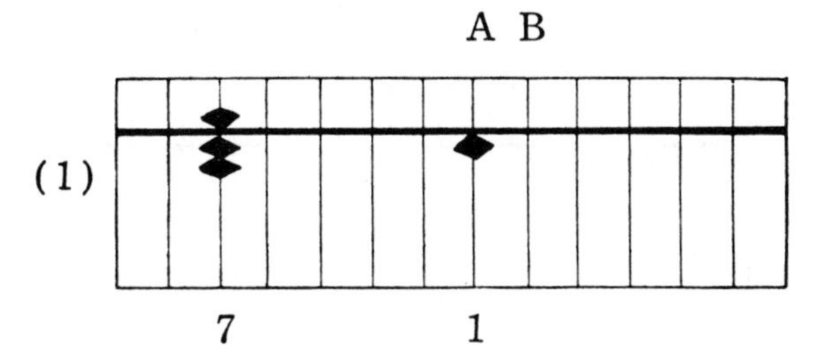

When we divide the 1 on the A upright by 7, we get, "7 into 1, 1 and 3 over." We raise three Earth counters on the next upright to the right of the A upright.

13 is indicated on the board.

2 and 6 over quotient
Divisor 7) 2 dividend

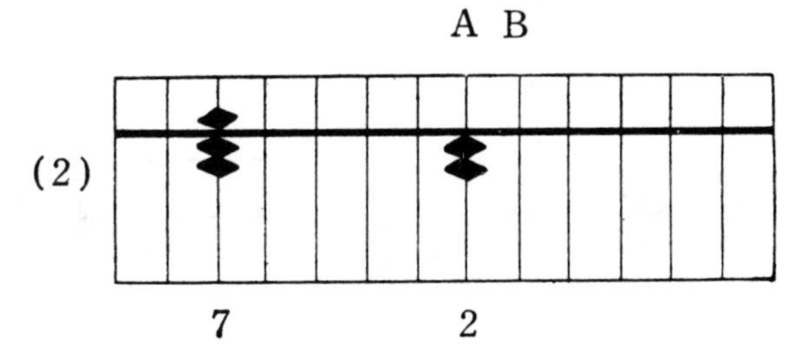

When we divide the 2 on the A upright by 7, we get, "7 into 2, 2 and 6 over." We raise 6 on the next upright to the right of the A upright.

26 is indicated on the board.

4 and 2 over quotient
Divisor 7) 3 dividend

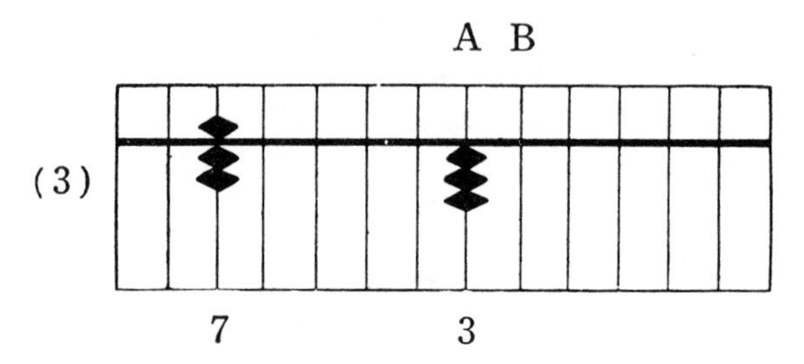

When we divide the three on the A upright by 7, we get, "7 into 3, 4 and 2 over." Instead of returning to place the three Earth counters on the A upright, add one Earth counter to these. Then raise two Earth counters on the next upright to the right, or the B upright.

42 is indicated on the board.

5 and 5 over quotient
Divisor 7) 4 dividend

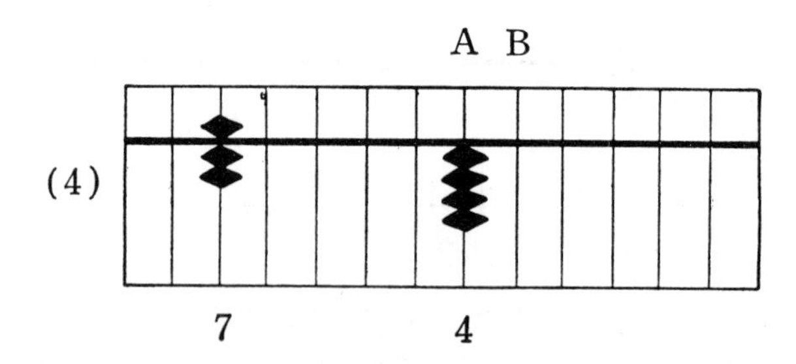

When we divide the 4 on the A upright by 7, we get, "7

into 4, 5 and 5 over." Instead of returning to place the four Earth counters on the A upright add one Earth counter to these and then slide down the Heaven counter on the next upright to the right, or the B upright.

55 is indicated on the board.

7 and 1 over quotient
Divisor 7) 5 dividend

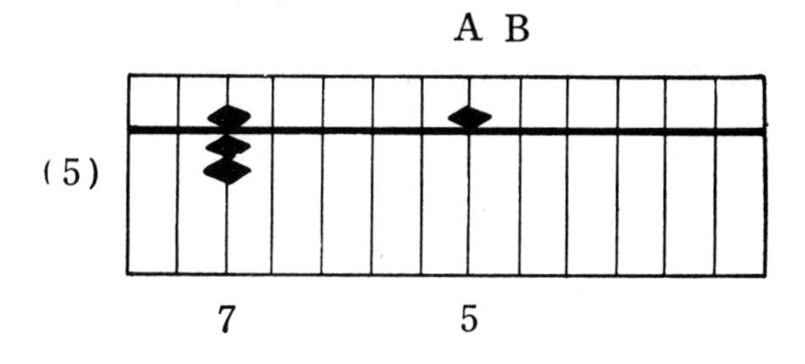

When we divide the 5 on the A upright by 7, we get, "7 into 5, 7 and 1 over." Instead of returning to place the Heaven counter on the A upright, add two Earth counters to it and then raise one Earth counter on the next upright to the right.

71 is indicated on the board.

8 and 4 over quotient
Divisor 7) 6 dividend

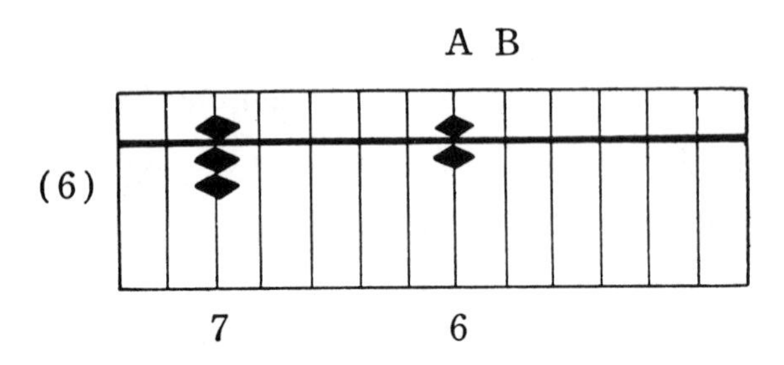

When we divide the 6 on the A upright by 7, we get, "7 into 6, 8 and 4 over." Instead of returning to place the counters which indicate 6 on the A upright, add two Earth counters to these and then raise four Earth counters on the next upright to the right, or the B upright.

84 is indicated on the board.

$$\text{Divisor}\quad 7\overline{)\,7\,}\ \text{dividend} \qquad \text{quotient: } 1$$

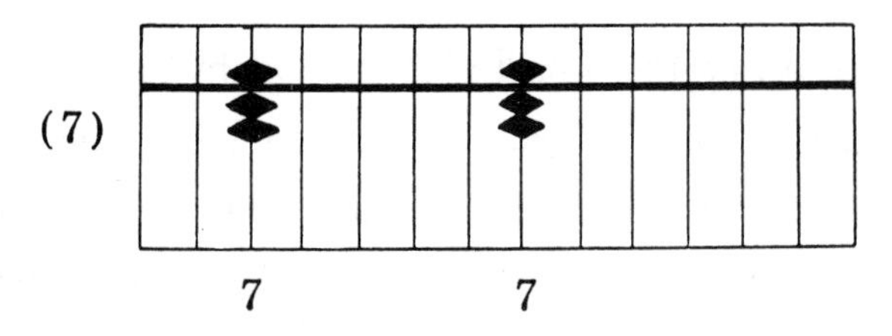

When we divide the 7 on the A upright by 7, we get, "7 into 7, advance 1." Return to place the counters which indicate 7 on the A upright and then raise one Earth counter on the next upright to the left.

1 is indicated on the board.

Dividing by 8.

8 into 1	1 and 2 over
8 into 2	2 and 4 over
8 into 3	3 and 6 over
8 into 4	Heaven 5
8 into 5	6 and 2 over
8 into 6	7 and 4 over
8 into 7	8 and 6 over
8 into 8	advance 1

Illustrations of Division by 8.

1 and 2 over quotient
Divisor 8) 1 dividend

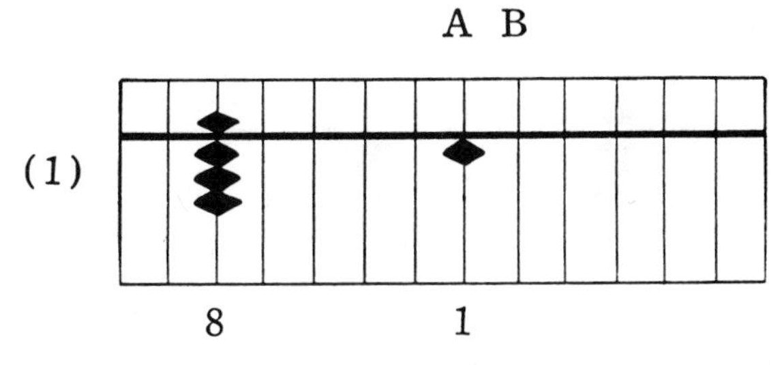

When we divide the 1 on the A upright by 8, we get, "8 into 1, 1 and 2 over." We raise two Earth counters on the next upright to the right of the A upright, or the B upright.

12 is indicated on the board.

2 and 4 over quotient
Divisor 8) 2 dividend

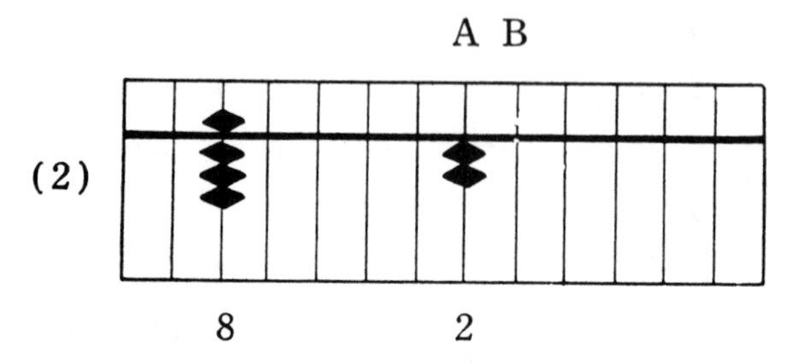

When we divide the 2 on the A upright by 8, we get, "8 into 2, 2 and 4 over." Raise four Earth counters on the next upright to the right of the A upright, or the B upright.

24 is indicated on the board.

3 and 6 over quotient
Divisor 8) 3 dividend

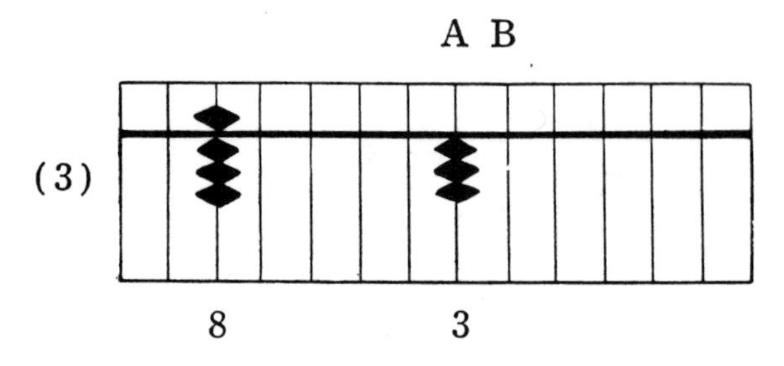

When we divide the 3 on the A upright by 8, we get, "8 into 3, 3 and 6 over." We indicate 6 on the next upright to theright of the A upright, or the B upright.

36 is indicated on the board.

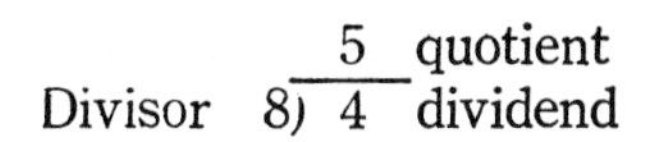

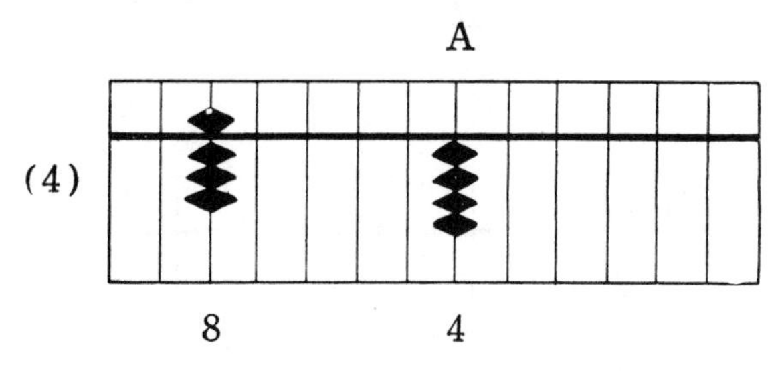

When we divide the 4 on the A upright by 8, we get, "8 into 4, Heaven 5." Return to place the four Earth counters on the A upright and slide down the Heaven counter on the same upright.

5 is indicated on the board.

6 and 2 over quotient
Divisor 8) 5 dividend

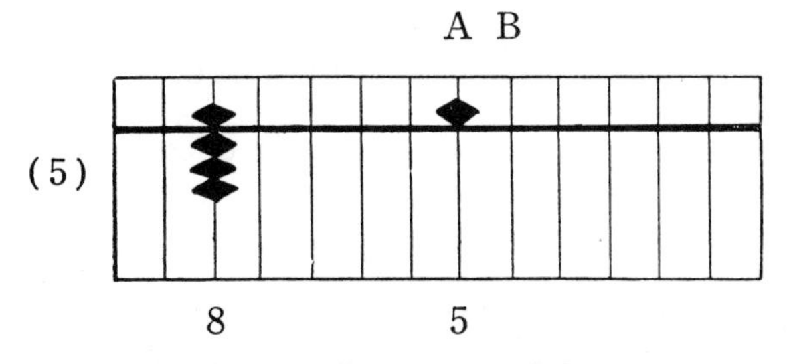

When we divide the 5 on the A upright by 8, we get, "8 into 5, 6 and 2 over." Instead of returning to place the Heaven counter on the A upright, add one Earth counter to it and raise two Earth counters on the next upright to the right, or the B upright.

62 is indicated on the board.

$$\text{Divisor}\quad 8\overline{)\,6\,}\ \text{dividend} \qquad \text{7 and 4 over quotient}$$

A B

(6)

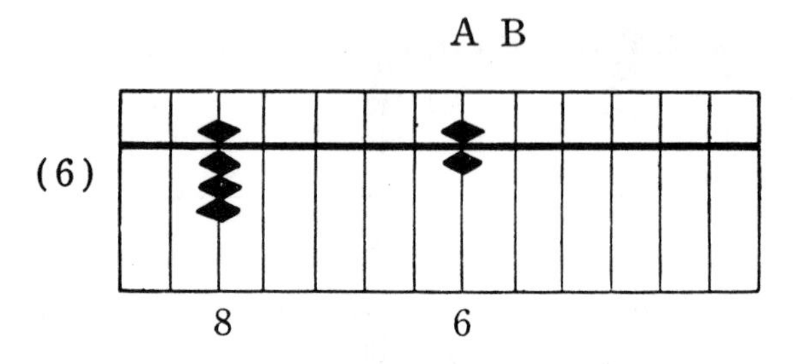

8 6

When we divide the 6 on the A upright by 8, we get, "8 into 6, 7 and 4 over." Instead of returning to place the counters which indicate 6 on the A upright, add one Earth counter to them and then raise four Earth counters on the next upright to the right, or the B upright.

74 is indicated on the board.

$$\text{Divisor}\quad 8\overline{)\,7\,}\ \text{dividend} \qquad \text{8 and 6 over quotient}$$

A B

(7)

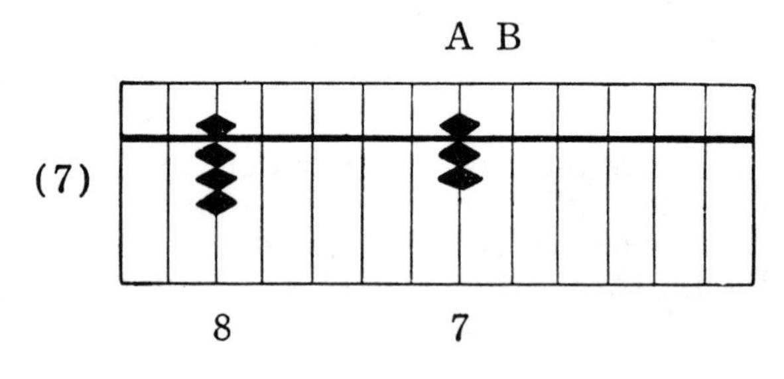

8 7

When we divide the 7 on the A upright by 8, we get, "8 into 7, 8 and 6 over." Instead of returning to place the counters which indicate 7 on the A upright, add one Earth counter to them and on the next upright to the right, indicate 6.

86 is indicated on the board.

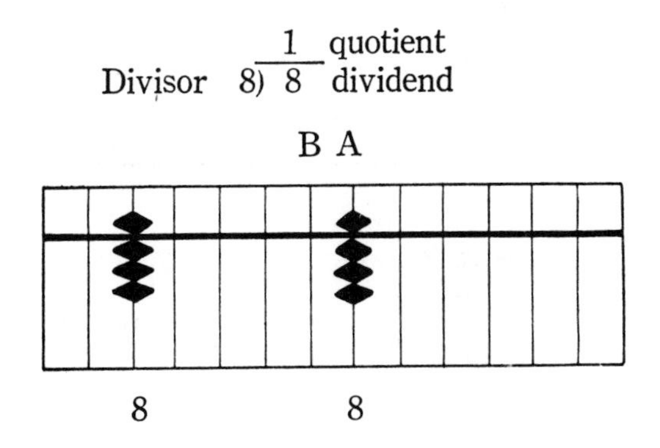

When we divide the 8 on the A upright by 8, we get, "8 into 8, advance 1." Return to place the counters which indicate 8 on the A upright, and raise one Earth counter on the next upright to the left.

1 is indicated on the board.

Dividing by 9.

9	into	1	1 and	1	over
9	into	2	2 and	2	over
9	into	3	3 and	3	over
9	into	4	4 and	4	over
9	into	5	5 and	5	over
9	into	6	6 and	6	over
9	into	7	7 and	7	over
9	into	8	8 and	8	over
9	into	9	advance 1		

Illustrations of Division by 9.

1 and 1 over quotient
Divisor 9) 1 divipend

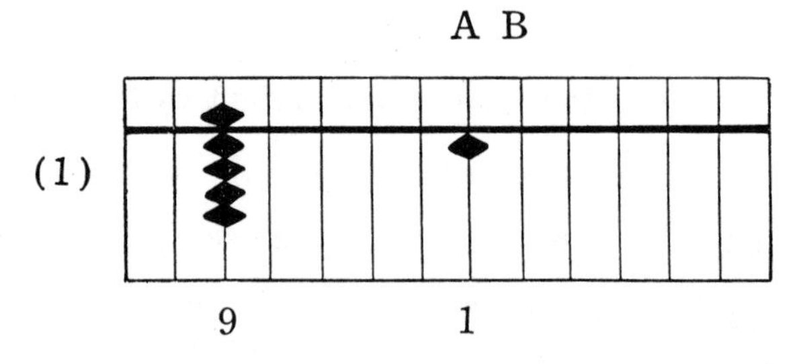

When we divide the 1 on the A upright by 9, we get, "9 into 1, 1 and 1 over." Raise one Earth counter on the next upright to the right, or the B upright.

11 is indicated on the board.

9 into 1 through 9 into 8 are solved by the same method.

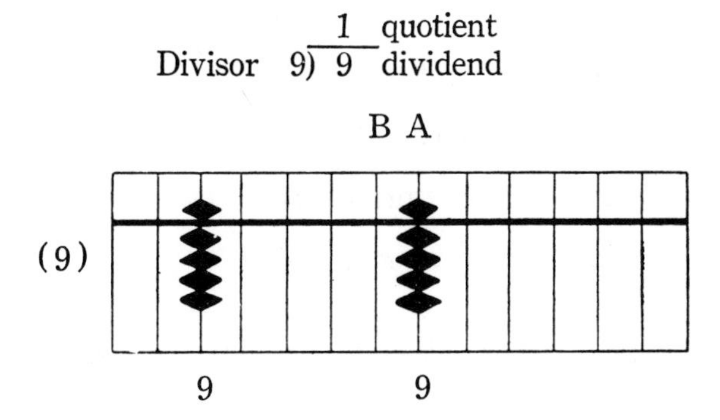

When we divide the 9 on the A upright by 9, we get, "9 into 9, advance 1." Return to place the counters which indicate 9 on the A upright and raise one Earth counter on the next upright to the left, or the B upright.

1 is indicated on the board.

Practical Application of the Divison Table.

2 into 4	advance 2
2 into 6	advance 3
2 into 8	advance 4
3 into 6	advance 2
3 into 9	advance 3
4 into 8	advance 2

These division tables should not be difficult to learn by heart.

Long division in writing is complicated and difficult to do accurately, but if you learn the Division Tables, long division is very simple on the abacus, because it is accomplished by dividing the figures on each upright one by one, and therefore, the figures are always between 1 and 9.

Practical Use of the Tables in "Division by Unification."

Memorization of the tables is absolutely necessary before trying to divide by this method.

If the first figure of the divisor is 1, we use the table for dividing by 1. If the first figure of the divisor is 2 we use the table for dividing by 2.

The following example will show how to proceed by this method.

Example 1.

In this case the first figure of the dividend is larger than the first figure of the divisor.

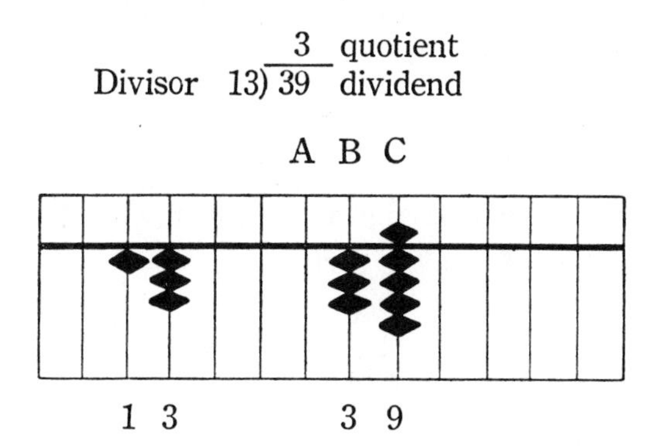

As the first figure of the divisor is 1, we must use the table for dividing by 1.

(1) Dividing the 3, which is the first figure of the dividend, by the 1, which is the first figure of the divisor, we get, "1 into 3 advance 3." Return to place the three Earth Counters on the first upright of the dividend and raise 3 Earth counters on the next upright to the left, in this case the A upright. 309 should be indicated on the board.

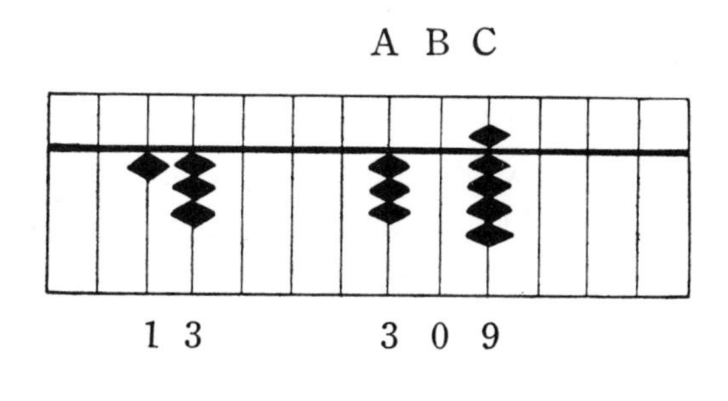

Now we multiply the 3 of the temporary quotient on the A upright by the 3 which is the second figure of the divisor. This gives us, "3×3 are 9." We then return to place the counters which indicate 9 on the last upright of the dividend.

3 is indicated on the board, and it is the quotient.

Example 2.

In this case, the first figure of the dividend is smaller than the first figure of the divisor.

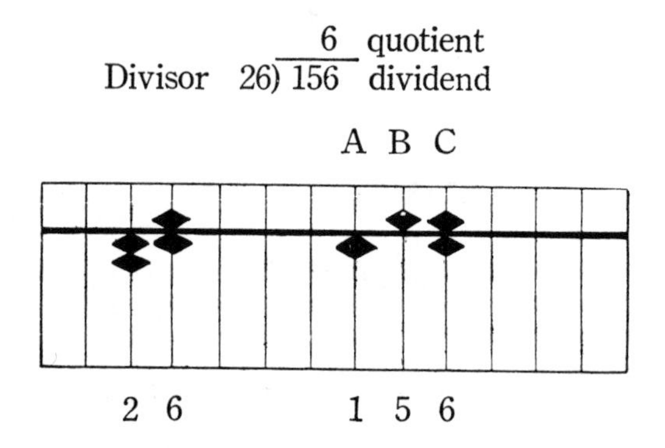

As the first figure of the divisor is 2, we must use the table for dividing by 2.

(1) Dividing the 1, which is the first figure of the dividend, by the 2, which is the first figure of the divisor, we get, "2 into 1 Heaven 5." Return to place the Earth counter on the first upright of the dividend, and then slide down the Heaven counter on the same upright. 556 should now be indicated on the board.

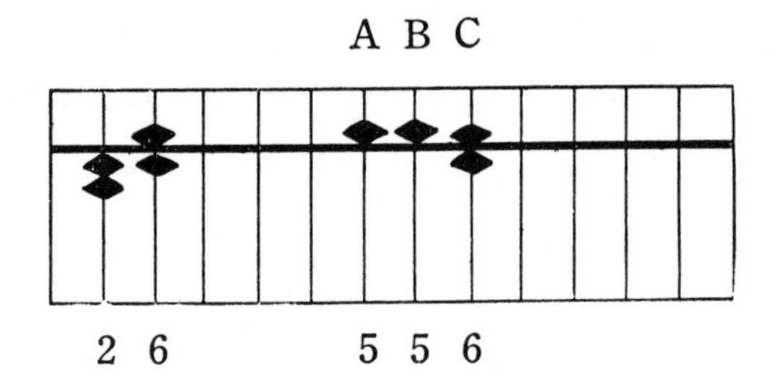

(2) Dividing the 5, which is the second figure of the dividend, by the 2, which is the first figure of the divisor, we get, "2 into 2 advance 1." (explained later.) Return to place the two Earth counters on the upright for tens in the dividend, and raise one Earth counter on the next upright to the left. As there is already a Heaven counter in place on that upright, we now have 6 there, and the board should indicate 636.

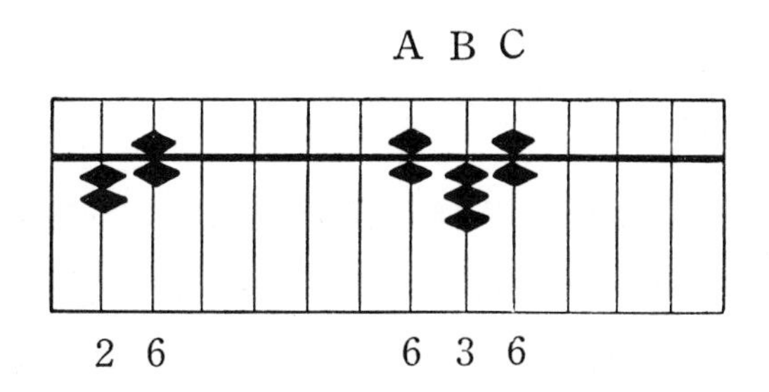

(3) Multiplying the 6 of the temporary quotient, which is on the A upright by the 6 on the end upright of the divisor,

we get, "6×6 are 36." We return to place the three Earth counters on the upright for tens and the counters which indicate 6 on the end upright of the dividend.

6 is indicated on the board. It is the quotient.

Please note: in spite of the fact that when we divide the 5 on the upright for tens by 2, we get 2 in the temporary quotient, we say, "2 into 2 advance 1." Otherwise, we would get into trouble, for if we said, "2 into 4 advance 2", and we indicated 2 on the first upright of the dividend, 716 would be indicated on the board.

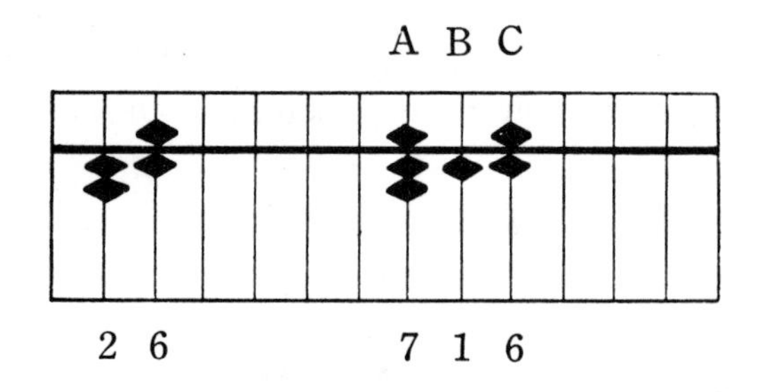

Then multiplying the 7 of the temporary quotient on the A upright of the dividend by the 6 on the end upright of the divisor, we would get, "6×7 are 42." But we could not subtract 42 from 16. Therefore, we must use, "2 into 2 advance 1."

Example 3.

6 quotient
Divlsor 36) 216 dividend

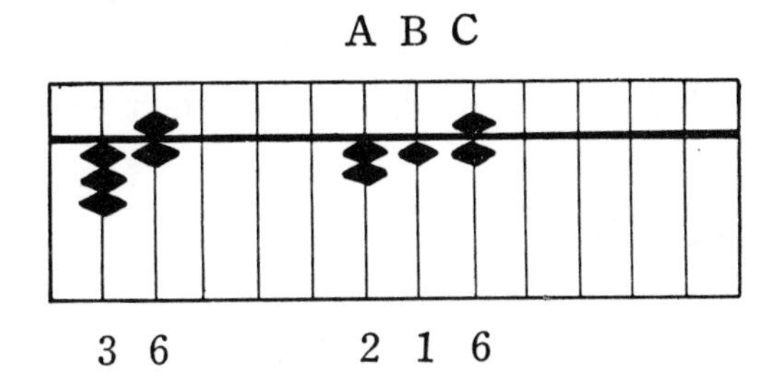

As the first figure of the divisor is 3, we must use the table for divididg by 3.

(1) Dividing the 2 on the first upright of the dividend by the 3 on the first upright of the divisor, we get, "3 into 2 6 and 2 over." Instead of returning to place the 2 Earth counters on the first upright of the dividend, add 4 to this same upright and raise two Earth counters on the next upright to the right, or the B upright. 636 should be indicated on the board.

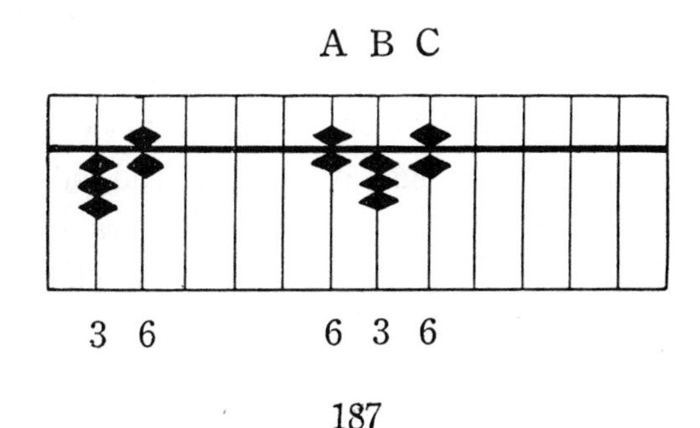

Multiplying the 6 of the temporary quotient on the first upright of the dividend, by the 6 on the last upright of the divisor, we get, "6×6 are 36." Return to place the three Earth counters on the upright for tens in the dividend and the 6 on the last upright of the dividend.

6 is indicated on the board. It is the quotient.

Example 4.

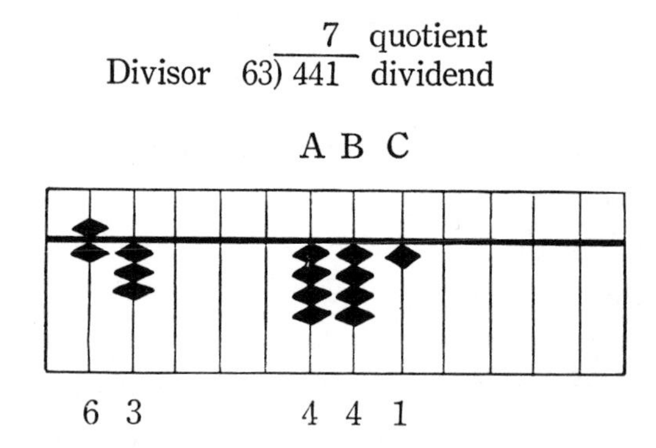

As the first figure of the divisor is 6, we must use the table for dividing by 6.

(1) Dividing the 4 on the first upright of the dividend by the 6 on the first upright of the divisor, we get, "6 into 4 6 and 4 over." Instead of returning to place the 4 Earth counters on the first upright of the dividend, add two to the same upright, and raise four Earth counters on the next upright to the right, or the B upright. 681 should be indicated on the board.

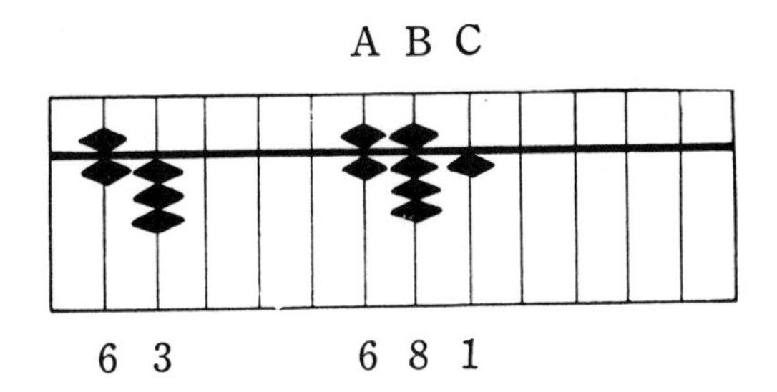

(2) Dividing the 8 on the B upright by 6, we get, "6 into 6 advance 1." We return to place the counters which indicate 6 of the 8 indicated on the B upright, and then raise one Earth counter on the next upright to the left. 721 should be indicated on the board.

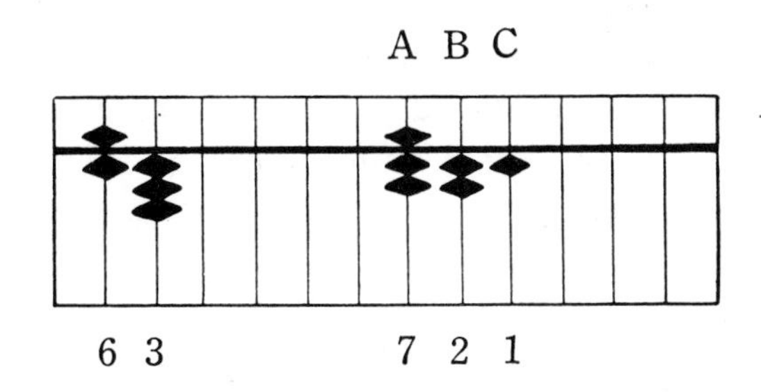

(3) Multiplying the 7 of the temporary quotient on the first upright of the dividend and the 3 on the last upright of the divisor, we get, "3×7 are 21." Return to place the two Earth counters on the upright for tens in the dividend and one Earth counter on the unit upright.

7 is indicated on the board. It is the quotient.

Example 5.

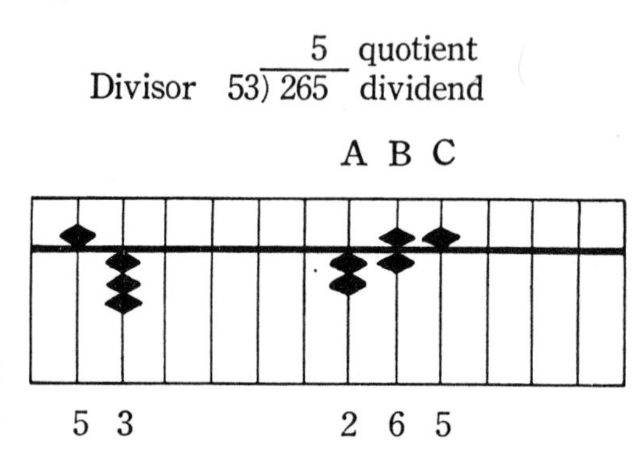

As the first figure of the divisor is 5, we must use the table for dividing by 5.

(1) Dividing the 2, which is the first figure of the dividend, by 5, on the first upright of the divisor, we get, "5 into 2 plus 2." Raise two Earth counters on the first upright of the dividend. 465 should be indicated on the board.

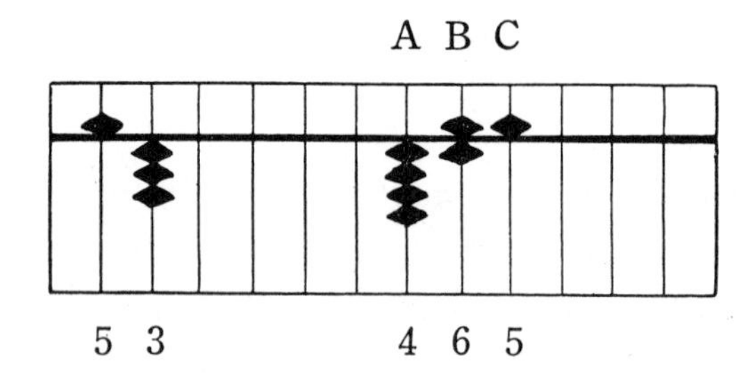

(2) Dividing the 6 on the B upright of the dividend by the 5 on the first upright of the divisor, we get, "5 into 5 advance 1." Return to place the Heaven counter on the B

upright of the dividend and raise one Earth counter on the next upright to the left, or the A upright. 515 should be indicated on the board.

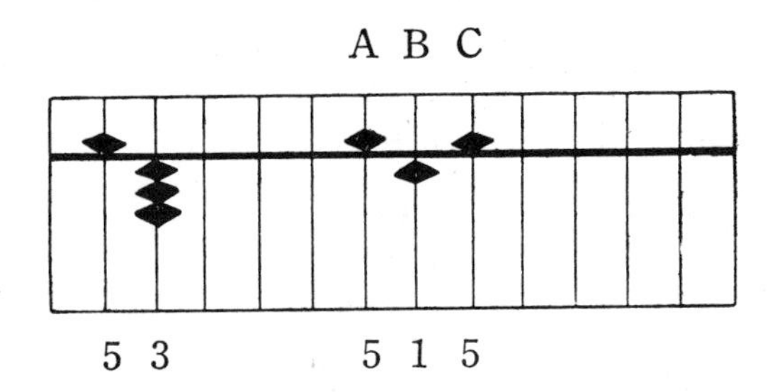

(3) Multiplying the 5 of the temporary quotient on the first upright of the dividend and the 3 on the last upright of the divisor, we get, "3×5 are 15." Return to place the Earth counter on the upright for tens in the dividend and the Heaven counter on the unit upright.

5 is indicated on the board. It is the quotient.

Example 6.

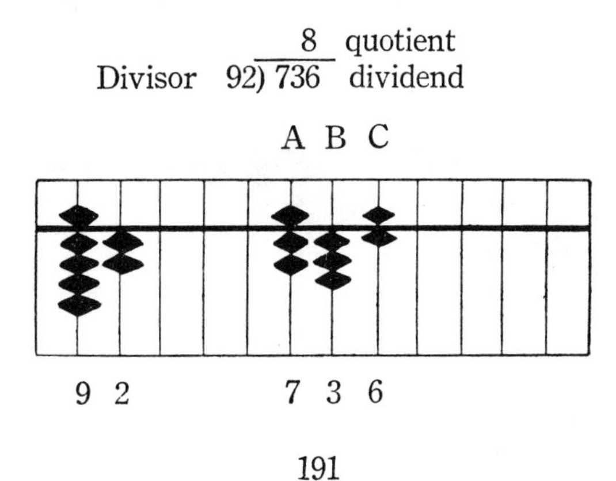

As the first figure of the divisor is 9, we must use the table for dividing by 9.

(1) Dividing the 7 on the first upright of the dividend by the 9 on the first upright of the divisor, we get, "9 into 7, 7 and 7 over." We then wish to put 7 on the B upright, but there are already three Earth counters in position there. If we add the 7 to the 3, the upright for tens will be full, therefore, we must divide the 10 by the 9 on the first upright of the divisor. This gives us, "9 into 9 advance 1." Instead of adding the 7 to the 3 on the upright for tens of the dividend, we return to place the two Earth counters on the same upright. Raise one Earth counter on the next upright to the left, or the A upright.

816 should be indicated on the board.

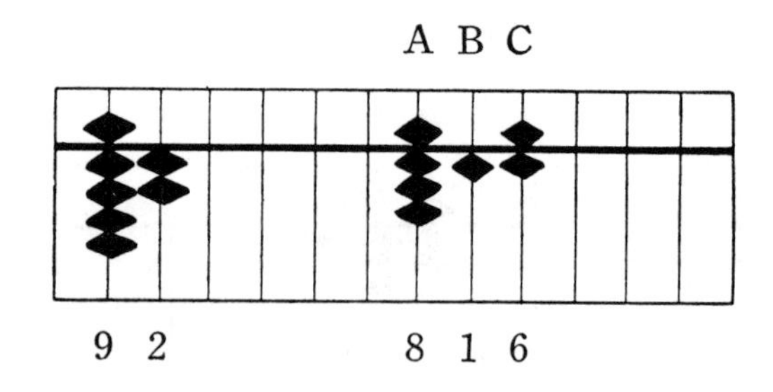

(2) Multiplying the 8 of the temporary quotient on the first upright of the dividend and the 2 on the last upright of the divisor, we get, "2×8 are 16." Return to place the Earth counter on the upright for tens in the dividend and the counters, which indicate 6 on the unit upright.

8 is indicated on the board. It is the quotient.

Please note that sometimes in the process of division an upright will become full as it did in the previous example, or you may have more than 10 to indicate on one upright. In such cases we must think how much we should add to the number in order to make it the same as the divisor, and then subtract that number from the lower upright and raise one Earth counter on the next upright to the left, as is shown in the following illustration.

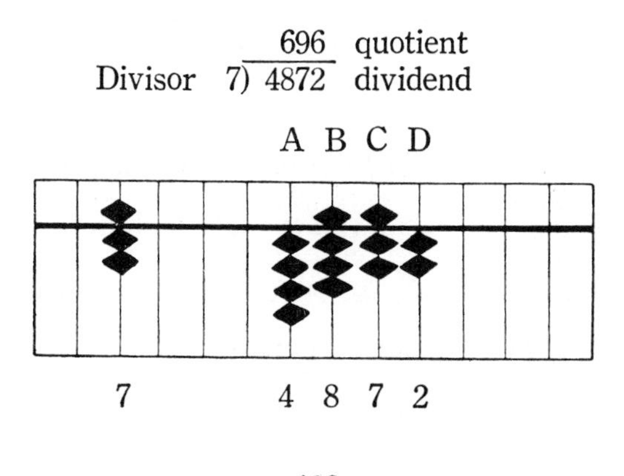

Dividing the 4 on the first upright of the dividend by the divisor, we get, "7 into 4, 5 and 5 over." Slide down the Heaven counter on the first upright of the dividend and return to place the four Earth counters on the same upright and add 5 on the next upright to the right. But there are already counters indicating 8 on that upright, and if we add 5 to it, there would be more than 10. Therefore, we keep the 5 in mind. 5872 would be indicated on the board, but with the 5 in our minds it would mean 5 (13) 72.

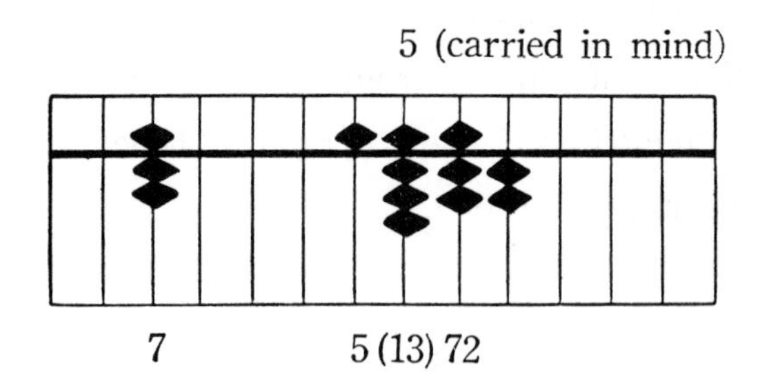

(2) Dividing the 13 on the upright for hundreds by the divisor, we get, "7 into 7 advance 1." Remembering the extra 5, we return to place only 2 Earth counters on that upright and raise one Earth counter on the next upright to the left. The board should now indicate 6672.

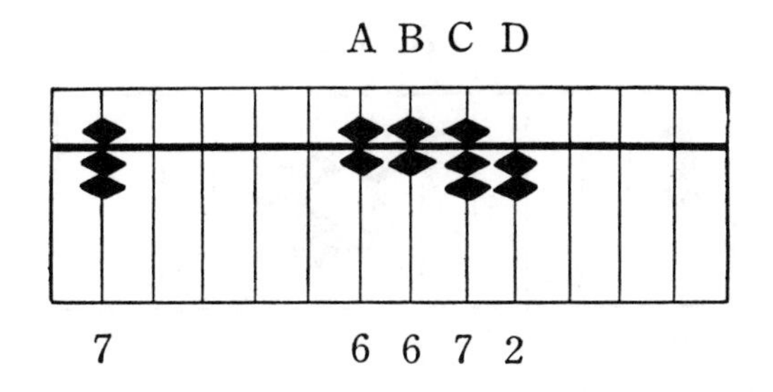

(3) Dividing the 6 on the B upright of the dividend by the divisor, we get, "7 into 6, 8 and 4 over." Instead of returning to place the counters which indicate 6 on B upright, add two Earth counters to them and raise four Earth counters on the next upright to the right. But this next step is impossible, as 7 is already indicated on that upright, therefore, we carry the 4 in our minds. The board indicates 6872 but with what we are carrying in our minds, indicates 68 (11) 2.

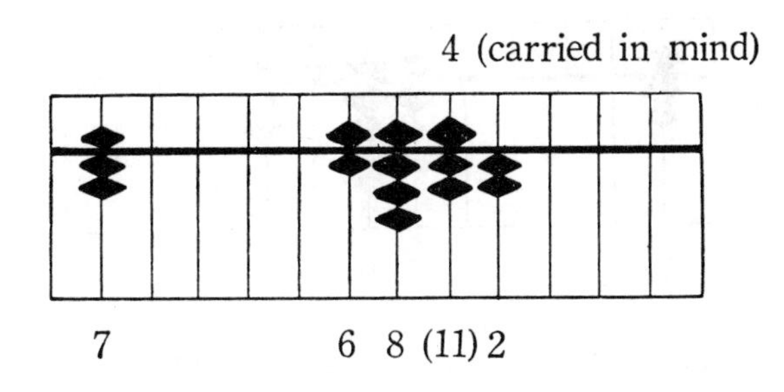

(4) As we could not add the four Earth counters to the 7, we must subtract 3 from the 7 indicated on the upright for tens in the dividend, and then raise one Earth counter on the next upright to the left, as to divide by 7, we, "7, into 7 advance 1." The board should now indicate 6942.

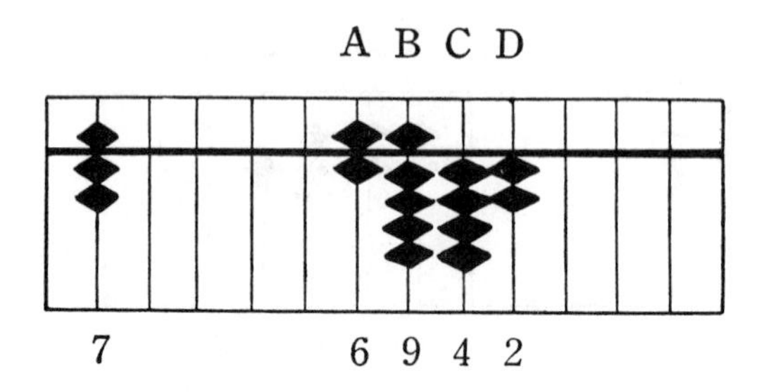

(5) Dividing the 4 on the C upright by the divisor, we get, "7 into 4, 5 and 5 over." Slide down the Heaven counter on the C upright, and at the same time return to place the four Earth counters on the same upright. Then slide down the Heaven counter on the next upright to the right. The board should now indicate 6957.

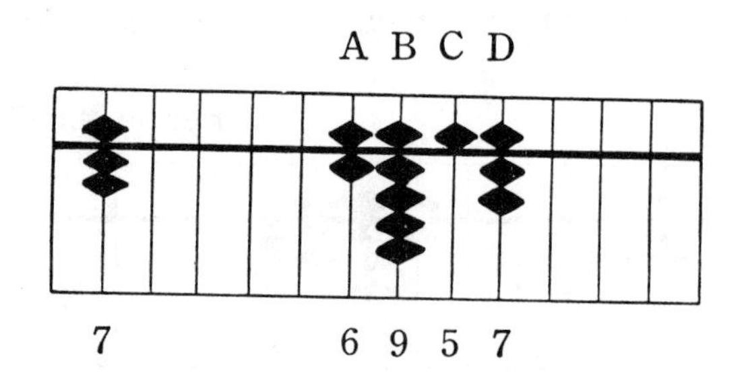

(6) Dividing the 7 on the D upright by the divisor, we get, "7 into 7 advance 1." Return to place the counters which indicate 7 on the D upright and raise one Earth counter on the next upright to the left, or the C upright.

696 is indicated on the board. It is the quotient.

Part 3.

Division Table by Transfiguration.

When the first figure of the divisor is the same as the first figure of the dividend and the second figure of the divisor is larger than the second figure of the dividend, there is another method of division used. This is called Division by Transfiguration, that is a change of figures.

I will show you this method by examples.

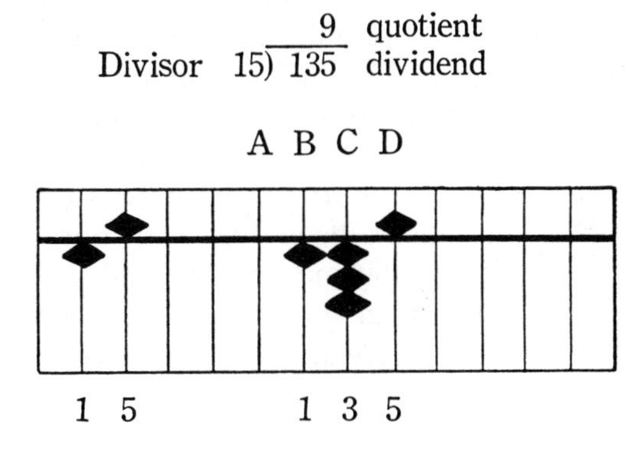

(1) Dividing the 1 on the first upright of the dividend by the 1 on the first upright of the divisor, we get, "1 into 1 advance 1." Return to place the Earth counter on the first upright of the dividend and raise one Earth counter on the next upright to the left.

1035 should be indicated on the board.

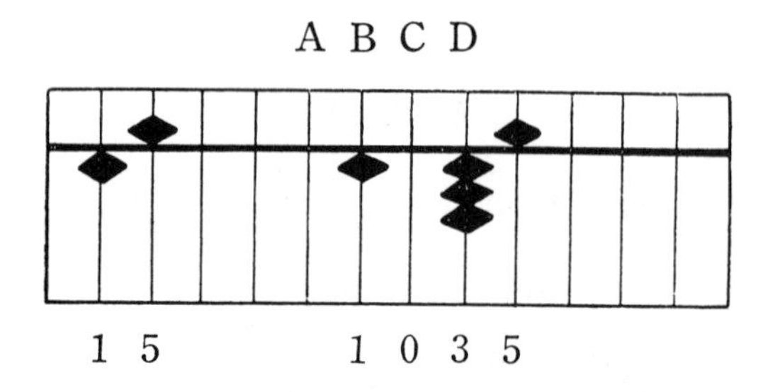

(2) If we multiply the 1 of the temporary quotient on the A upright by the 5 on the end upright of the divisor, we get, "1 × 5 is 5." But we cannot subtract this 5 from the 3 on the upright for tens in the dividend, and having no other resource, we must take back the Earth counter which we advanced to the A upright. In such cases as this we must use the Division Table by Transfiguration.

"looking at the 1, indicate 91 without dividing."

(3) Return to place the Earth counter on the A upright of the dividend and indicate 9 on the next upright to the right, or the upright for 100's in the dividend, and raise one Earth counter on the next upright to the right, or the upright for 10's in the dividend. 945 should be indicated on the board.

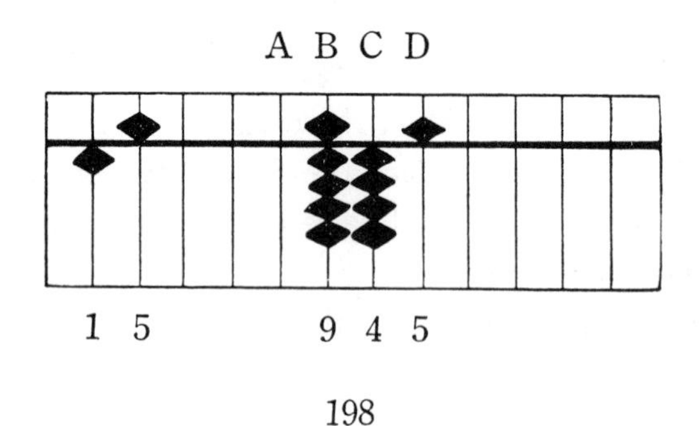

(4) Multiplying the 9 of the temporary quotient on the first upright of the dividend by the 5 on the last upright of the divisor, we get, "9×5 are 45." Return to place the four Earth counters on the upright for 10's in the dividend and the Heaven counter on the last upright.

9 is indicated on the board. It is the quotient.

As we mentioned above, there is a Division Table to be used when we cannot use the ordinary Division Table. It is as follows:

Rule for 1.	Looking at the 1, indicate 91 without dividing
Rule for 2,	Looking at the 2, indicate 92 without dividing
Rule for 3,	Looking at the 3, indicate 93 without dividing
Rule for 4,	Looking at the 4, indicate 94 without dividing
Rule for 5,	Looking at the 5, indicate 95 without dividing
Rule for 6.	Looking at the 6, indicate 96 without dividing
Rule for 7.	Looking at the 7, indicate 97 without dividing

Rule for 8.	Looking at the 8, indicate 98 without dividing
Rule for 9.	Looking at the 9, indicate 99 without dividing

Part 4.

Practical use of the Division Table by Transfiguration.

Example 1.

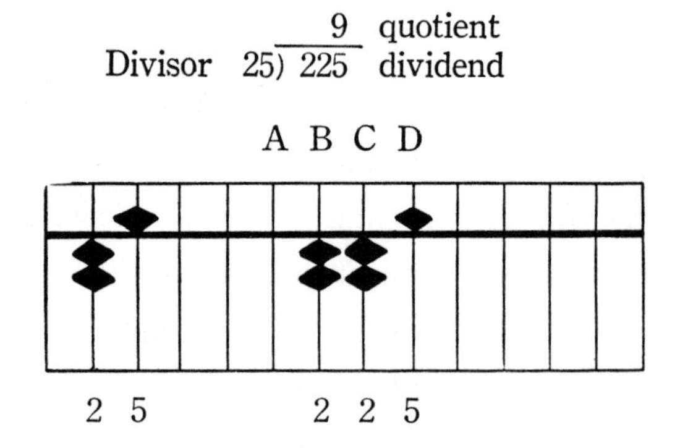

As the 2 on the first upright of the divisor is the same as the first figure of the dividend, and the 5 on the second upright of the divisor is larger than the 2 on the second upright of the dividend, we must use the rule for 2.

(1) Dividing the 2 on the first upright of the dividend by the 2 on the first upright of the divisor, we get, "2 into 2

advance 1." Return to place the two Earth counters on the first upright of the dividend and raise one Earth counter on the next upright to the left, or the A upright. The board should now indicate 1025.

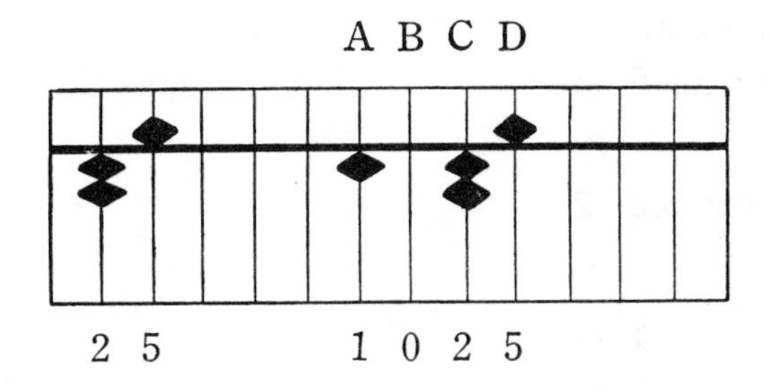

(2). If we multiply the 5 on the last upright of the divisor by the 1 of the temporary quotient on the A upright of the dividend, we get, "1×5 is 5," but we cannot subtract this 5 from the 2 on the C upright of the dividend, so we must take back the 2 which was advanced to the A upright, and using the Division Table for 2, we get, "looking at the 2, indicate 92 without dividing." We now change the 2 on the B upright to 9, and add two Earth counters on the next upright to the right, or the C upright. The board should now indicate 945.

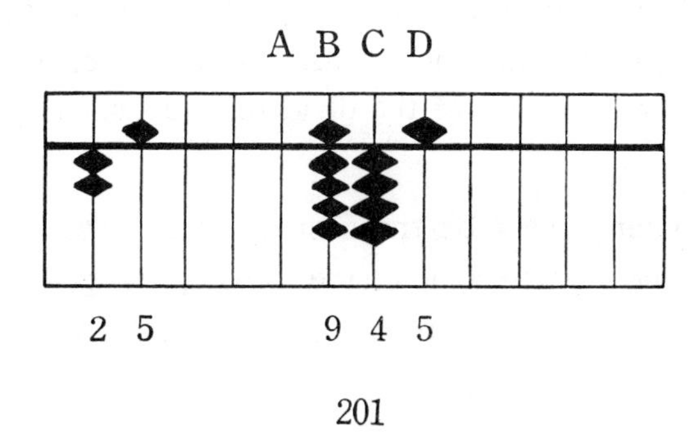

Multiplying the 9 of the temporary quotient on the B upright of the dividend and the 5 on the last upright of the divisor, we get, "5×9 are 45." Return to place the four Earth counters on the C upright and the Heaven counter on the D upright.

9 is indicated on the board. It is the quotient.

Example 2.

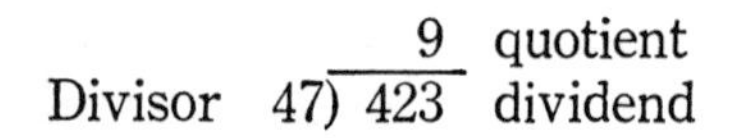

$$\begin{array}{rr} & 9 \\ \text{Divisor} \quad 47) & \overline{423} \end{array} \quad \begin{array}{l} \text{quotient} \\ \text{dividend} \end{array}$$

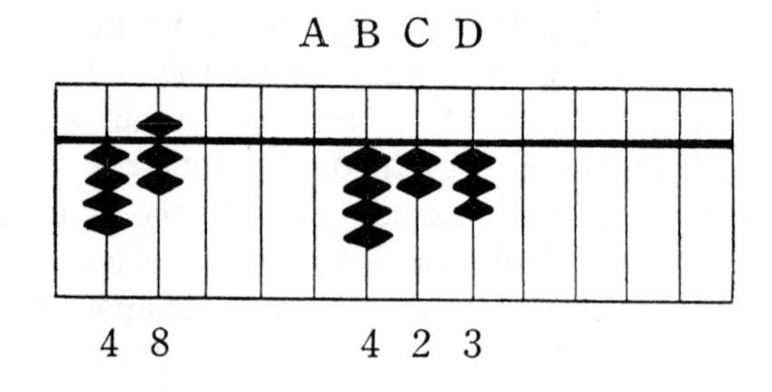

As the 4 on the first upright of the divisor is the same as the 4 on the first upright of the dividend, and the 7 on the second upright of the divisor is larger than the 2 on the second upright of the dividend, we must use the rule for 4.

(1) Dividing the 4 on the first upright of the dividend by the 4 on the first upright of the divisor, we get, "4 into 4 advance 1." Return to place the four Earth counters on the

first upright of the dividend and raise one Earth counter on the next upright to the left, or the A upright. 1023 should be indicated on the board.

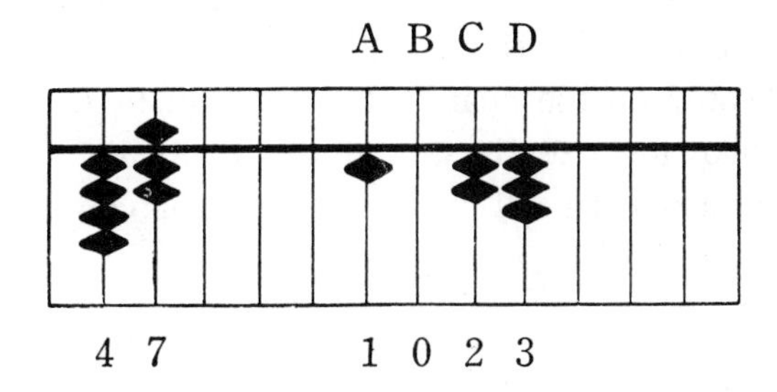

(2) If we multiply the 1 of the temporary quotient on the A upright of the dividend by the 7 on the last upright of the divisor, we get, "1 × 7 is 7", but we cannot subtract the 7 from the 2 on the C upright of the dividend, therefore, we must take back the 4 counters which were advanced to the A upright and using the rule for 4, we get, "looking at the 4, indicate 94 without dividing." Changing the 4 to 9 and adding four Earth counters on the next upright to the right, we should get 963 indicated on the board.

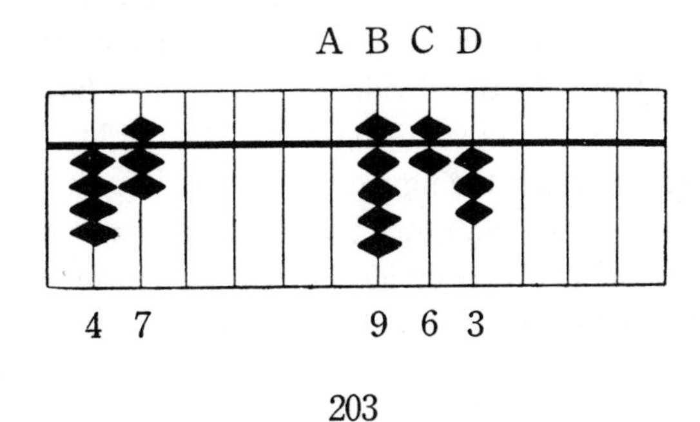

(3) Multiplying the 9 of the temporary quotient on the first upright of the dividend and the 7 on the last upright of the divisor, we get, "7×9 are 63." Return to place the counters which indicate 6 on the C upright and the three Earth counters on the D upright.

9 is indicated on the board. It is the quotient.

In the following exercises, find the quotient. Cover the answers with a card. If your work is correct, there will be no remainders.

$25\,\overline{)\,2425}$	$34\,\overline{)\,3128}$	$86\,\overline{)\,8428}$
Use rule for 2	Use rule for 3	Use rule for 8
Answer 97	Answer 92	Answer 98

Part 5.

Division Table by Replacement.

This division table is used in the following ways. After having divided the first figure of the dividend by the first figure of the divisor, and multiplying the first figure of the temporary quotient by the second figure of the divisor, we find that we cannot subtract the product from the figure on the lower upright of the dividend, so we must borrow 1 or 2 and place it on the next upright to the right. Then we multiply again, using the figure which was left after subtracting, and the product can now be subtracted from the figure on the lower upright of the dividend. In such cases we use the Division Table by Replacement. I will show you by example.

3 quotient

Divisor 15) 45 dividend

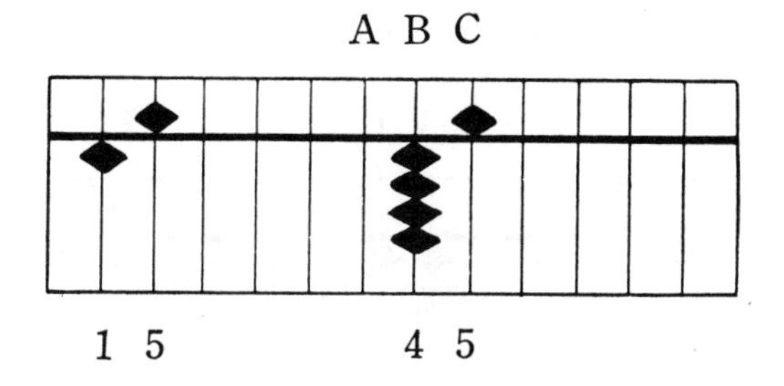

(1) Dividing the 4 on the first upright of the dividend by the 1 on the first upright of the divisor, we get, "1 into 4 advance 4." Return to place the four Earth counters on the first upright of the dividend and raise four Earth counters on the next upright to the left, or the A upright. The board should indicate 105.

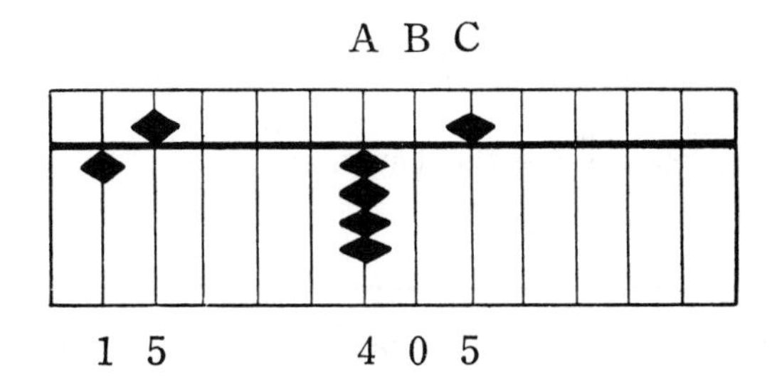

(2) Multiplying the 4 of the temporary quotient on the first upright of the dividend by the 5 on the last upright of the divisor, we get, "4×5 are 20", but we cannot subtract the 20 from the 5 on the C upright of the dividend, so we must use the Division Table by Replacement.

(3) We take one Earth counter from the 4 on the A upright of the temporary quotient and place it on the B upright. 315 will then be indicated on the board.

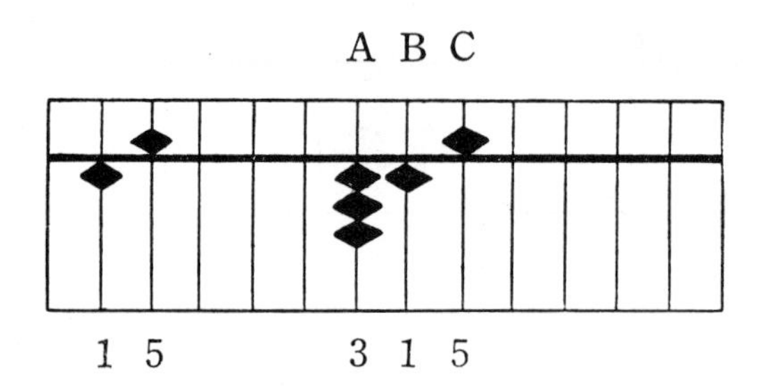

(4) Multiplying the 3 on the A upright of the temporary quotient by the 5 on the last upright of the divisor, we get, "3×5 are 15." Now, return to place the Earth counter on the B upright of the dividend and the Heaven counter on the C upright.

3 is indicated on the board. It is the quotient.

In such cases we use the Division Table by Replacement, which is as follows:

Rule for 1 (When we have 1 on the first upright of the divisor)

"Subtract 1, place 1 on the next right hand upright."

Rule for 2 (When we have 2 on the first upright of the divisor)

"Subtract 1, place 2 on the next right hand upright."

Rule for 3 (When we have 3 on the first upright of the divisor)

"Subtract 1, place 3 on the next right hand upright."

Rule for 4 (When we have 4 on the first upright of the divisor)

"Subtract 1, place 4 on the next right hand upright."

Rule for 5 (When we have 5 on the first upright of the divisor)

"Subtract 1, place 5 on the next right hand upright."

Rule for 6 (When we have 6 on the first upright of the divisor)

"Subtract 1, place 6 on the next right hand upright."

Rule for 7 (When we have 7 on the first upright of the divisor)

"Subtract 1, place 7 on the next right hand upright."

Rule for 8 (When we have 8 on the first upright of the divisor)

"Subtract 1, place 8 on the next right hand upright."

Rule for 9 (When we have 9 on the first upright of the divisor)

"Subtract 1, place 9 on the next right hand upright."

Part 6.

Practical use of the Division Table by Replacement.

Example 1.

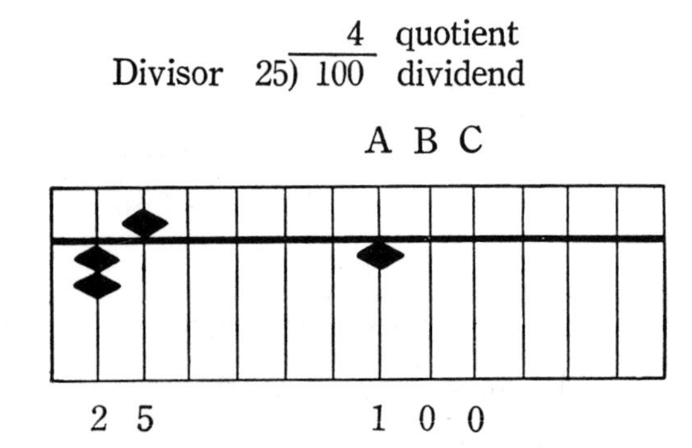

As we have 2 on the first upright of the divisor, we must use the rule for 2.

(1) Dividing the 1 on the first upright of the dividend by the 2 on the first upright of the divisor, we get, "2 into 1 Heaven 5." Slide down the Heaven counter on the first upright of the dividend, and at the same time return to place the Earth counter on the same upright. The board should indicate 500.

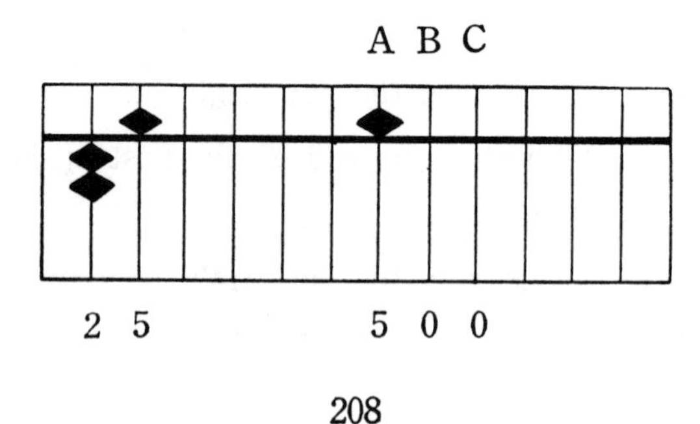

(2) Multiplying the 5 of the temporary quotient on the first upright of the dividend by the 5 on the last upright of the divisor, we get, "5×5 are 25", but we cannot subtract the 25 from the B and C uprights of the dividend, so we must use the rule for 2, "Subtract 1, place 2 on the next right hand uright." Subtracting 1 from the 5 indicated on the A upright, we get 4, so we return to place the Heaven counter and raise four Earth counters. Next, we raise two Earth counters on the next right hand upright, or the B upright. The board should then indicate 420.

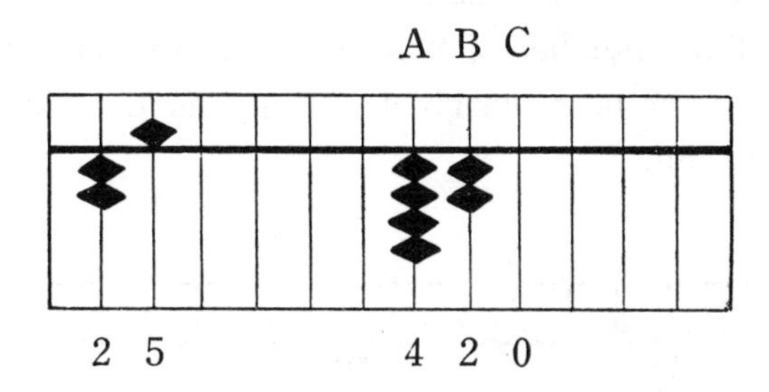

(3) Multiplying the 4 of the temporary quotient on the first upright of the dividend by the 5 on the last upright of the divisor, we get, "4×5 are 20." Return to place the two Earth counters on the B upright of the dividend.

4 is indicated on the board. It is the quotient.

To sum up the material we have given you, there are three Division Tables:

1. Division Table by Unification.
2. Division Table by Transfiguration.

3. Division Table by Replacement.

These are very important to you, if you wish to master division, for whether you gain speed or not they are very effective.

Section 2.

The Determination of Position.

(1) In cases when the divisor is a whole number:

Beginning with the unit position of the dividend, count off to the left the number of uprights that are contained in the divisor and the next upright to the left is the unit position of the quotient.

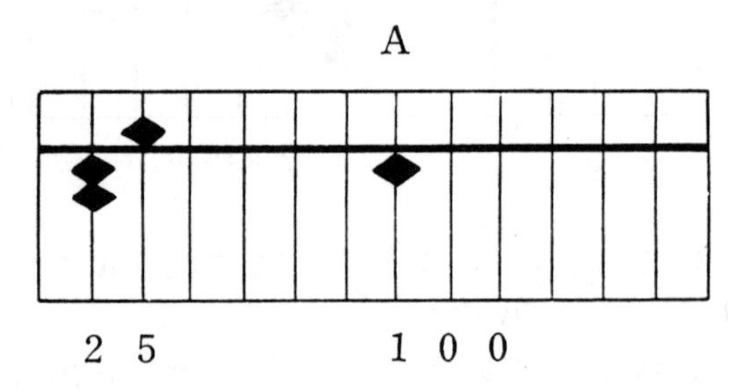

Divisor 25) $\overline{100}$ dividend

A upright is the unit position of the quotient.

(2) In cases when the divisor is a decimal fraction, such as 0.123:

If we have no zero between the decimal point and the first

figure of the fraction, the unit position of the dividend is the unit position of the quotient.

(3) In cases when the divisor is a decimal fraction with zeros to the right of the decimal, such as 0.00123:

Beginning with the unit position of the dividend, count off to the right the number of uprights as there are zeros to the right of the decimal point in the divisor, and the next upright to the right is the unit position for the quotient.

Divisor 0.00123) $\overline{456}$ dividend

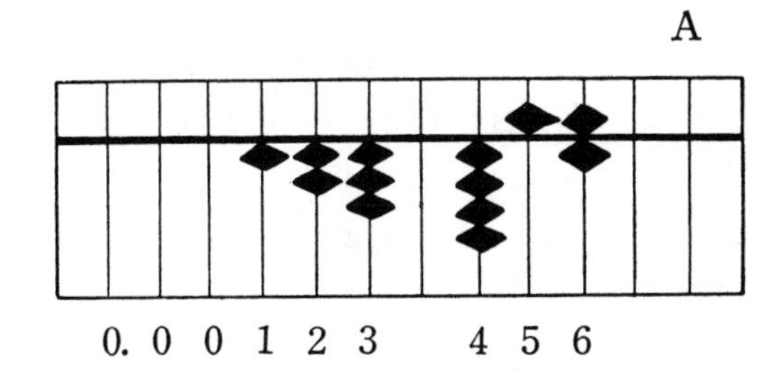

A upright is the unit position for the quotient.

Section 3.

The Process of Division.

(1) Indicate the dividend on the right hand side of the abacus, and the divisor on the left hand side.

(2) Look at the head figure of the dividend and the head figure of the divisor and get the temporary quotient and

determine which of the three Division Tables to use.

(3) Multiply the first figure of the temporary quotient by the second figure of the divisor and subtract the product from the lower uprights of the dividend.

(4) If the divisor contains many uprights, such as 23,456, we must multiply each number by the temporary quotient and subtract the product from the lower uprights of the dividend.

In such cases the unit position of the dividend from which the product is subtracted descends one position with each multiplication. The unit position of the product which we get by multiplying the second figure of the divisor and the first figure of the dividend becomes the position for tens in the next multiplication.

Continuing like this, we descend, one upright at a time, to the last upright of the dividend.

If we subtract the product from every upright of the dividend and still do not have the real quotient, we begin again by dividing the second upright of the dividend by the head upright of the divisor and continue as before.

(5) If we get the whole temporary quotient and we subtract all of the porducts which we get by multiplying every upright of the divisor by the temporary quotient, then we have the real quotient.

(6) If we cannot subtract the product which we get by multiplying the temporary quotient and the second upright of the divisor or any upright beneath it, from the second

upright of the dividend, we must use the Division Table by Replacement.

Part 1.

Practical Training in Division.

Example 1.

$$\begin{array}{lrl} & \underline{\quad 12} & \text{quotient} \\ \text{Divisor} & 56)\ 672 & \text{dividend} \end{array}$$

A B C D

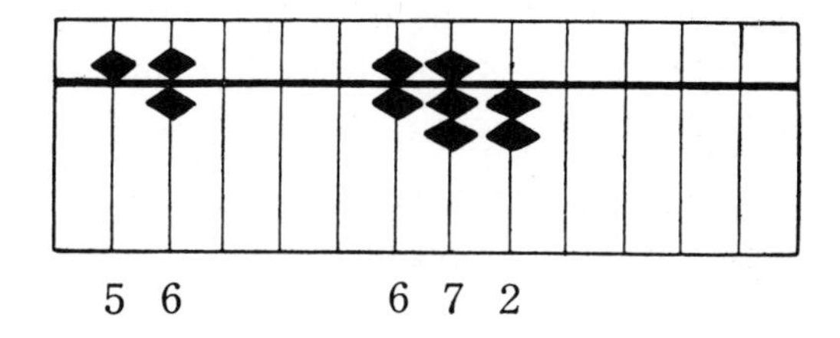

(1) Indicate the dividend on the right hand side of the board and the divisor on the left hand, as is shown in the above illustration.

(2) Count off the number of uprights which are contained in the divisor to the left, beginning with the unit position of the dividend, and the next left hand upright is the unit position for the quotient, In this case it is the B upright.

(3) Divide the first figure of the dividend by the first figure of the divisor, in this case, "5 into 5 advance 1." Return to place the Heaven counter on the first upright of the

dividend and raise one Earth counter on the next upright to the left, or the A upright. The board should indicate 1172.

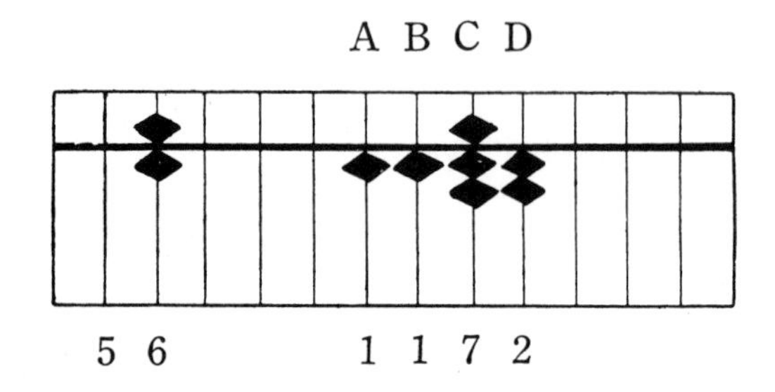

(4) Multiplying the 1 of the temporary quotient on the A upright of the dividend by the 6 on the end upright of the divisor, we get, "1×6 is 6." Subtracting this product from the 7 on the upright for tens in the dividend (In this case the B upright is the position for tens and the C upright is the unit position.), 1112 should be indicated on the board.

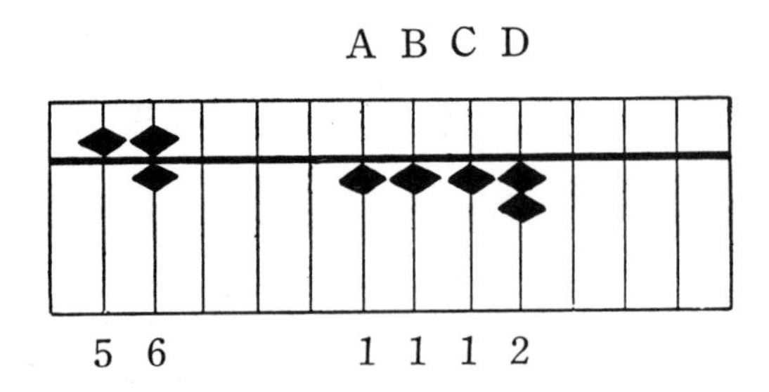

(5) Dividing the 1 on the B upright (not the 1 on the A upright, as it belongs to the temporary quotient) by the 5 on the first upright of the divisor, we get, "5 into 1 plus 1."

Raise one Earth counter on the upright for 100's in the dividend, or the B upright. 1212 should be indicated on the board.

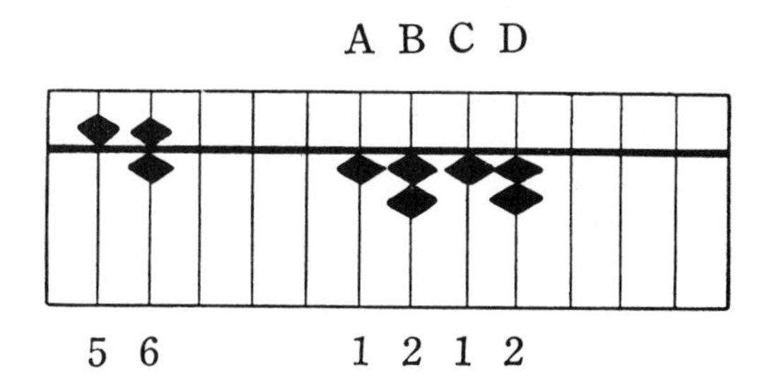

(6) Multiplying the 2 of the temporary quotient on the upright for 100's in the dividend by the 6 on the end upright of the divisor, we get, "2×6 are 12." Return to place the one Earth counter on the upright for 10's, or the C upright and the two Earth counters on the last upright of the dividend.

12 is indicated on the board. It is the quotient.

Example 2.

201 quotient, 118 remainder
Divisor 162) 32680 dividend

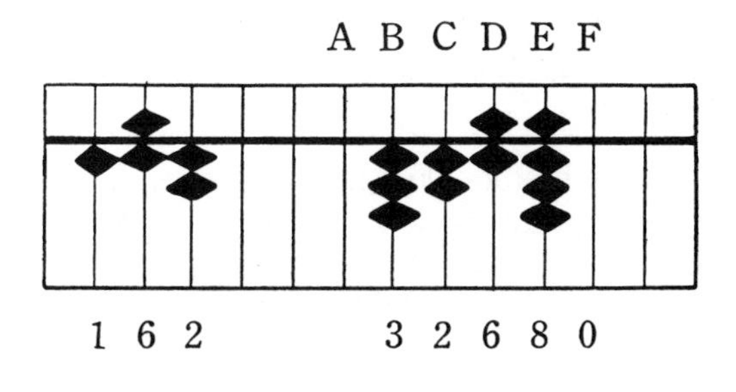

(1) Indicate the divisor on the left hand side of the board, and the dividend to the right of it, as is shown in the above illustration.

(2) Beginning with the unit position of the dividend, count off to the left the number of uprights which are contained in the divisor, and the next left upright is the unit position for the quotient.

(3) Dividing the 3 on the first upright of the dividend by the 1 on the first upright of the divisor, we get, "1 into 3 advance 3." Return to place the 3 Earth counters on the first upright of the dividend and at the same time raise three Earth counters on the next upright to the left, or the A upright. The board should indicate 302680.

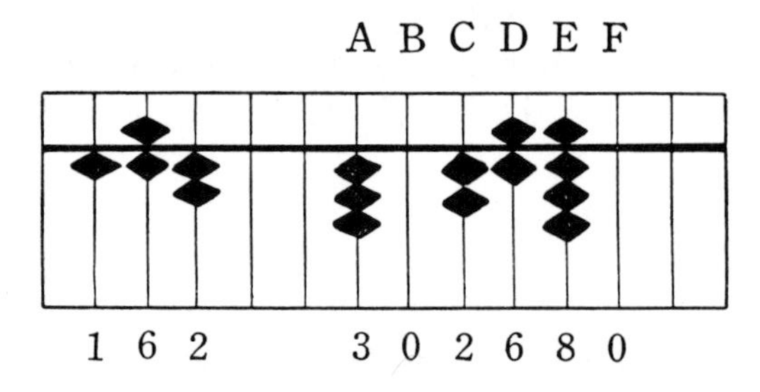

(4) Multiplying the 3 of the temporary quotient on the A upright by the 6 on the second upright of the divisor, we get, "3×6 are 18." But we cannot subtract the 18 from the 2 on the upright for 100's in the dividend.

(5) In this case, therefore, we must use the Division Table, "1 into 2 advance 2." Return to place the two Earth counters on the first upright of the dividend and at the same

time raise two Earth counters on the next upright to the left, or the A upright. The board should indicate 212680.

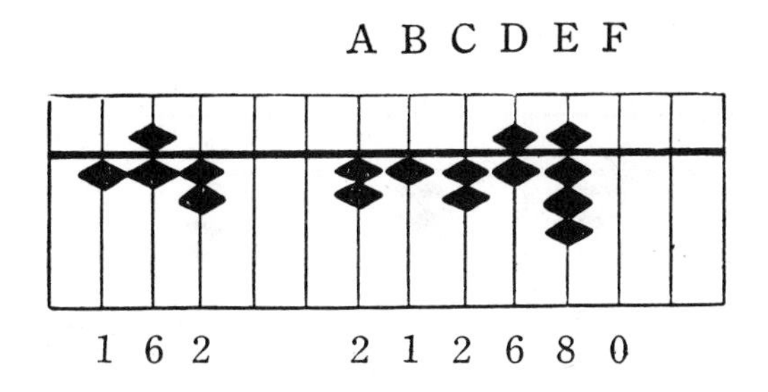

(6) Multiplying the 2 of the temporary quotient on the A upright by the 6 on the upright for 10's in the divisor, we get, "2×6 are 12." Return to place the one Earth counter on the upright for 10000's in the dividend and the two Earth counters on the upright for 1000's. The board should indicate 200680.

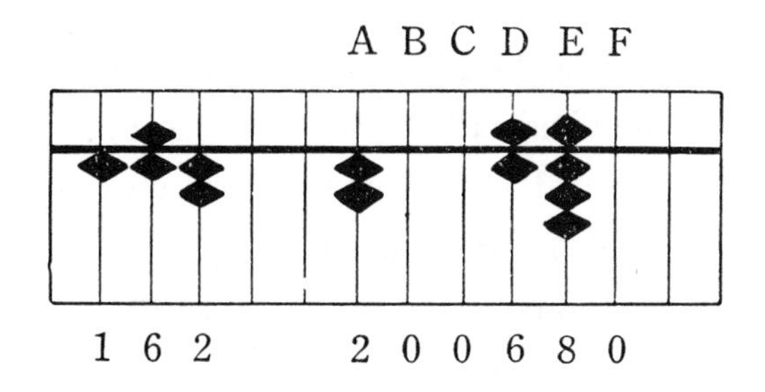

(7) Multiplying the 2 of the temporary quotient on the A upright by the 2 on the last upright of the divisor, we get, "2×2 are 4." Subtract 4 from the 6 indicated on the

upright for 100's of the dividend. The board should now indicate 200280.

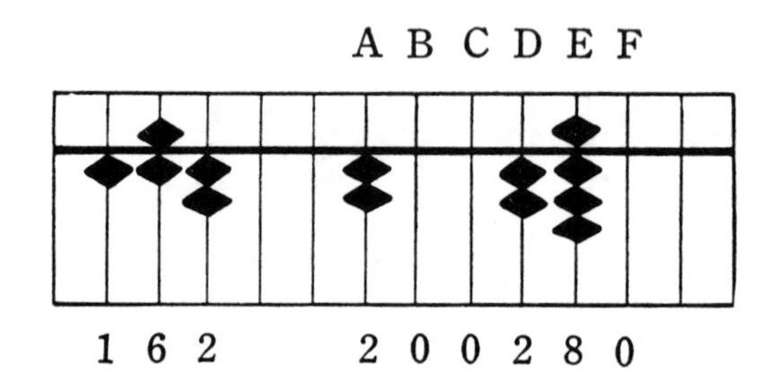

(8) Dividing the 2 on the upright for 100's in the dividend by the 1 on the first upright of the divisor, we get, "1 into 2 advance 2." But if we return to place the two Earth counters on the upright for 100's of the dividend and raise two Earth counters on the next upright to the left, we will not be able to subtract the product 12, which we get by multiplying the 2 of the temporary quotient on the upright for 1000's in the dividend by the 6 on the upright for 10's in the divisor, from the amount indicated on the upright for 100's and for 10's in the dividend.

Therefore, we must use the Division Table, "1 into 1 advance 1", and return to place one of the Earth counters on the upright for 100's in the dividend and raise one Earth counter on the next upright to the left, or the C upright. The board should indicate 201180.

A B C D E F

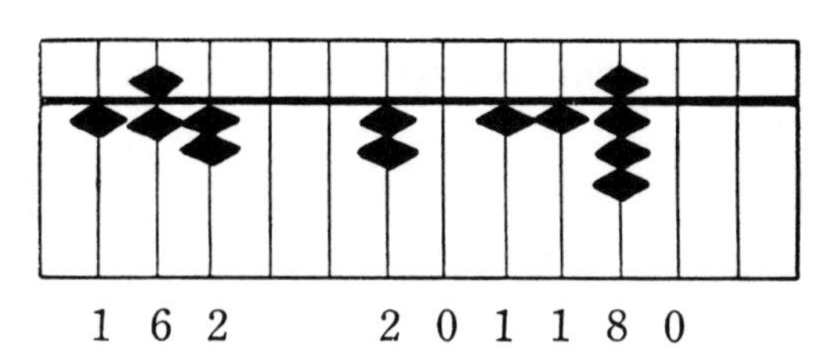

(9) Multiplying the 1 of the second temporary quotient on the upright for 1000's in the dividend by the 6 on the upright for 10's in the divisor, we get, "1×6 is 6," Return to place the counters which indicate 6 from the 8 on the upright for 10's in the dividend. The board should indidate 201120.

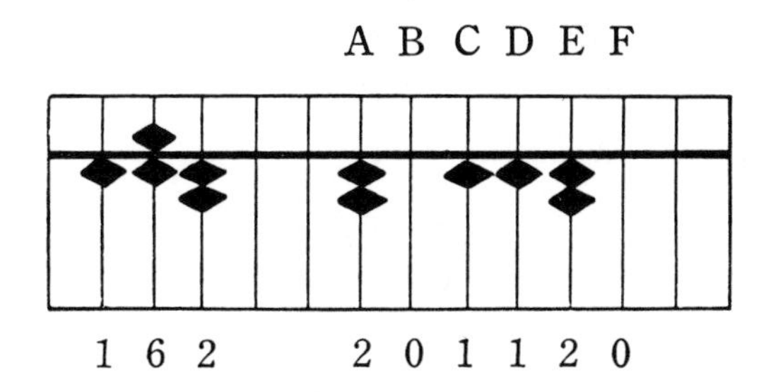

(10) Multiplying the 1 of the second temporary quotient on the upright for 1000's in the dividend by the 2 on the last upright of the divisor, we get, "1×2 is 2." We want to subtract this two from the last upright of the dividend, but as there is a zero there, we must think, "2 from 10 is 8." Then we return to place one Earth counter on the upright for 10's of the dividend and at the same time indicate 8 on the next upright to the right, or the last upright of the dividend. The board should now indicate 201118.

The C upright was the upright which we found to be the unit position of the quotient at the beginning of our division. Therefore, the 201 indicated on the A, B, and C uprights is the quotient, and the 118 on the D, E and F uprights is the remainder.

quotient	201
remainder	118

Part 2.

Find the quotients and the remainders in the following problems. If there is no remainder, write after the quotient, "and no remainder."

Check your answers by multiplying the quotient by the divisor and adding the remainder.

(1)	(2)	(3)
21) 466	36) 1278	25) 425
(4)	(5)	(6)
69) 2208	27) 513	23) 322

Find the quotients and check answers.

(1)	(2)	(3)
42) $ 63.84	52)$ 110.76	14) $ 71.12
(4)	(5)	(6)
16) $ 15.68	34) $ 38.08	27)$ 146.34

Practice examples for Division.

Find the quotient and the remainder.

(1) divisor $25\overline{)\,432}$ dividend

(2) divisor $108\overline{)\,1404}$ dividend

(3) divisor $98\overline{)\,2548}$ dividend

(4) divisor $27\overline{)\,244}$ dividend

(5) divisor $16\overline{)\,4032}$ dividend

(6) divisor $12\overline{)\,1668}$ dividend

(7) divisor $14\overline{)\,567}$ dividend

(8) divisor $54\overline{)\,1026}$ dividend

(9) divisor $36\overline{)\,972}$ dividend

(10) divisor $45\overline{)\,525}$ dividend

Key to the above problems.

(1) 17 and 7 over
(2) 13
(3) 26
(4) 9 and 1 over
(5) 252
(6) 139
(7) 40 and 7 over
(8) 19
(9) 27
(10) 11 and 30 over

CHAPTER VII.

DIVISION BY ABBREVIATION

This method of division uses the Multiplication Tables only and is like division in writing.

Section 1.

The Determination of Position.

(1) When the divisor is a whole number, count the number of uprights.

(2) Beginning with the unit position of the dividend, count off towards the left as many uprights as there are in the divisor, and the second next upright to the left is the unit position for the quotient. This is shown in the following illustration.

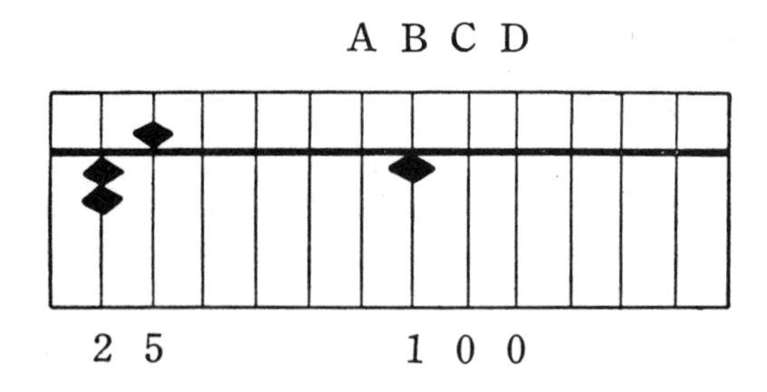

A upright is the unit position of the quotient.

If we compare the determination of the unit position for the quotient in Division by Abbreviation and Division by Unification, we will see that Division by Abbreviation we must count off one more upright to the left, than in Division by Unification.

Divisor 12) 345 dividend

(a) Division by Unification.

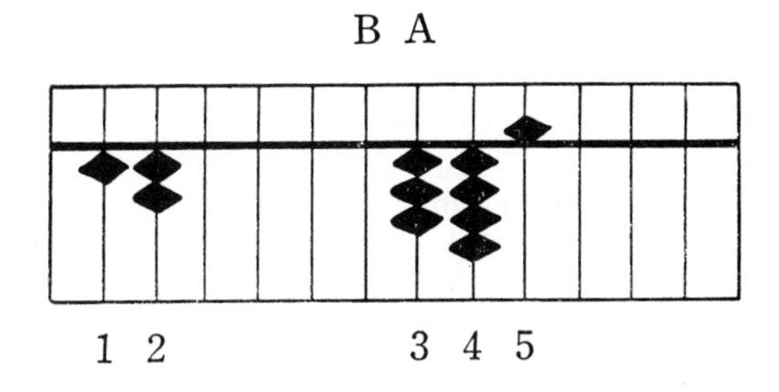

(b) Division by Abbreviation.

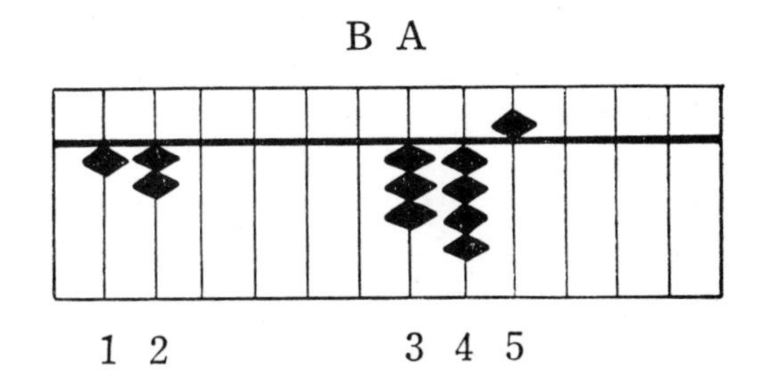

(a) A upright is the unit position.
(b) B upright is the unit position.

(3) In cases when the divisor is a decimal fraction such as 0.123 and there is no zero between the decimal point and the first figure, the next upright to the left of the unit position of the dividend is the unit position for the quotient.

(4) In cases when the divisor is a decimal fraction such as 0.00123, count the number of uprights which contain zero between the decimal point and the first figure and subtract 1 from this number, then count off this many uprights toward the right beginning with the unit position of the dividend and the next upright to the right is the unit position for the quotient.

Section 2.

The Process of Division by Abbreviation.

Type 1.

(1) Indicate the divisor on the left hand side of the abacus and the dividend to the right of it.

(2) We must divide the dividend into groups containing the same number of uprights as there are in the divisor. If we have a two place number for the divisor, we must divide the entire dividend into groups of two as is shown in the following illustration.

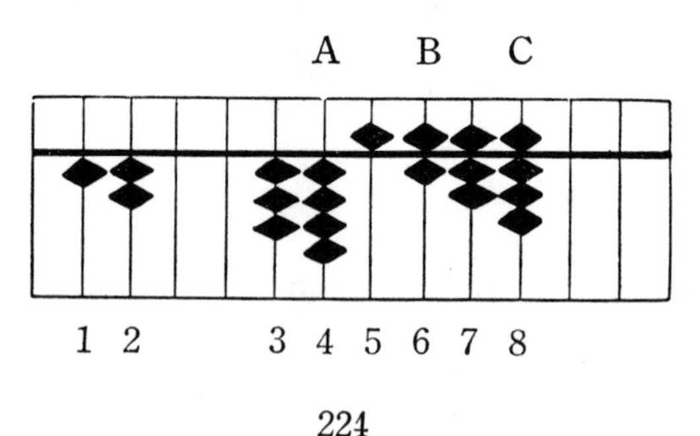

(3) Taking the first group only, we think of how many times the divisor is contained into this group.

(4) When we get the first temporary quotient in this way, we indicate it on the second upright to the left from the first upright of the dividend.

(5) Then we multiply the temporary quotient by the first upright of the divisor and subtract the product from the first number.

(6) In this way we must multiply every upright of the divisor by the first temporary quotient, and if we are able to subtract the product within the boundary of the first group, then we see how many times the divisor is contained in the second group.

(7) After we thus get the second temporary quotient, we indicate it on the next upright to the right of the first temporary quotient.

(8) Then we multiply the second temporary quotient by the first upright of the divisor and when we get the product, subtract it from the second group of the dividend.

(9) When we have proceeded in this way with every group of the dividend, we will have the real quotient and the remainder.

Section 3.

The Process of Division by Abbreviation.

Type 2.

(1) Indicate the divisor on the left hand side of the board and the dividend to the right of it.

(2) After we have divided the dividend into groups which have the same number of uprights as the divisor, we sometimes find that we cannot subtract some of the products from the number within the group, as is shown in the following:

$$\text{Divisor} \quad 20\overline{)\,21123\,} \quad \text{dividend}$$

As 20 is a two-place number, we divide the dividend into three groups, as follows: 21, 12, 3. But 20 will not go into the 12 of the second group, therefore, we must add one more upright and divide the dividend in the following way: 21, 123.

(3) In such a case we must be careful in placing the temporary quotient. It should be indicated on the next upright to the right of where it would ordinarily be placed.

When you multiply the temporary quotient by some upright of the divisor and get a number too large to subtract from the number in the group with which you are dealing, you proceed in just the same manner as you do when the number cannot be divided by the divisor.

(4) When the temporary quotient has three places or more, it is difficult to figure in our minds how many times the divisor of two places is contained in it. It will help us to consider the first figure of the dividend and see how many times the first figure of the divisor will go into it. For example, when you are dividing 72 by 21, think how many times 2 goes into 7.

Section 4.

Practical Training in Division by Abbreviation

Example 1.

$$\begin{array}{rrl} & 24 & \text{quotient} \\ \text{Divisor} \quad 3) & \overline{72} & \text{dividend} \end{array}$$

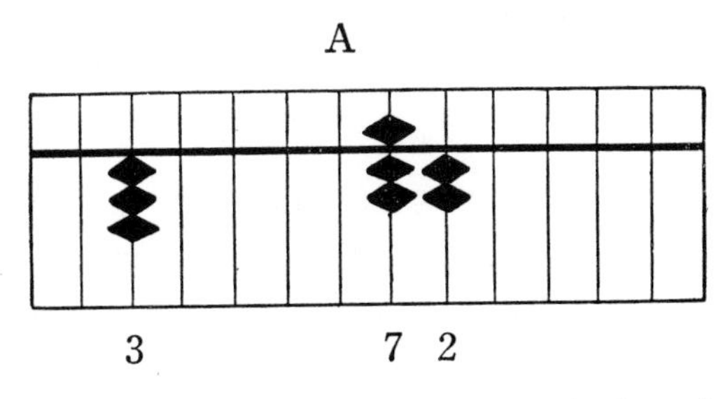

(1) Indicate the divisor on the left of the abacus and the dividend to the right of it, as is shown in the above illustration.

(2) Beginning with the unit position of the dividend, count off to the left the number of uprights which are contained in the divisor and the second upright to the left of that one is the unit position for the quotient.

(3) As the divisor is a one-place number, we must divide the dividend into groups containing one upright.

(4) By looking at the board we see that we must divide the 7 on the first upright of the dividend by the divisor which is 3.

(5) As 3 goes into 7 two times, we must raise two Earth counters on the second upright to the left of the first upright of the dividend. The board should indicate 2072.

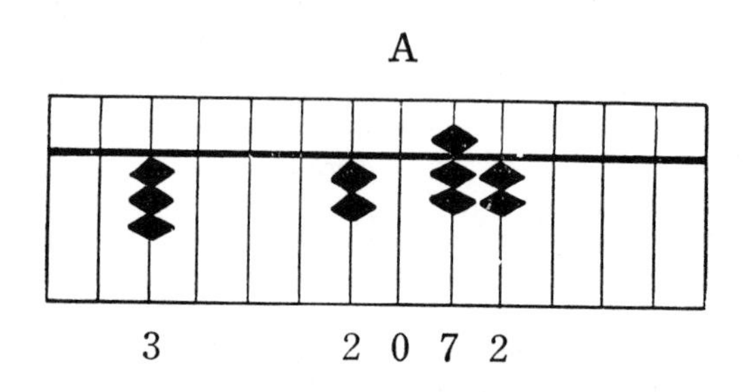

(6) Multiplying the 2 of the temporary quotient by the 3 of the divisor, we get, "2×3 are 6." Return to place 6 of the 7 indicated on the upright for 10's in the dividend. The board should now indicate 2012.

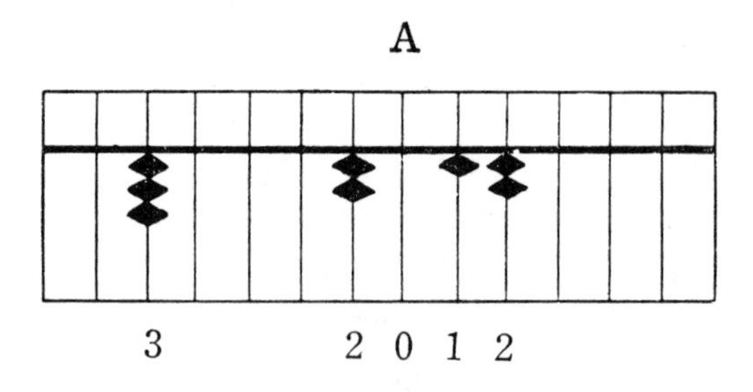

(7) As the divisor 3 will not go into the 1 on the upright for 10's in the dividend, we must add one more upright to it, then 3 will go into 12, 4 times.

(8) We must indicate this 4 on the next upright to the

right of the first temporary quotient. The board should indicate 2412.

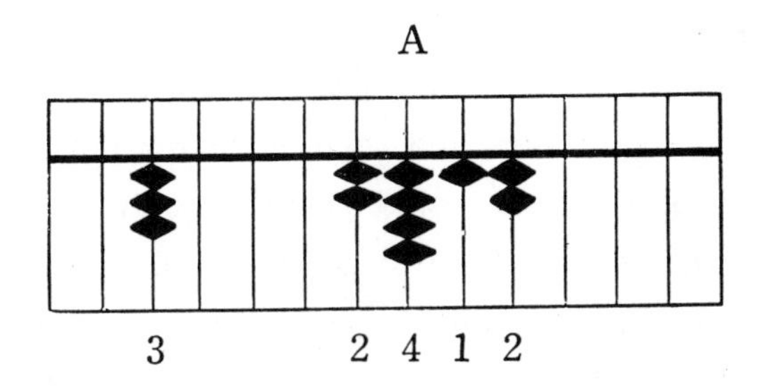

(9) Multiplying the 4 of the second temporary quotient by the divisor, we get, "3×4 are 12." Return to place the Earth counter on the upright for 10's in the dividend and the two Earth counters on the last upright.

24 is indicated on the board. It is the quotient.

Example 2.

$$\begin{array}{rr@{}l} & 14 & \text{ quotient} \\ \text{Divisor } 26) & \overline{364} & \text{ dividend} \end{array}$$

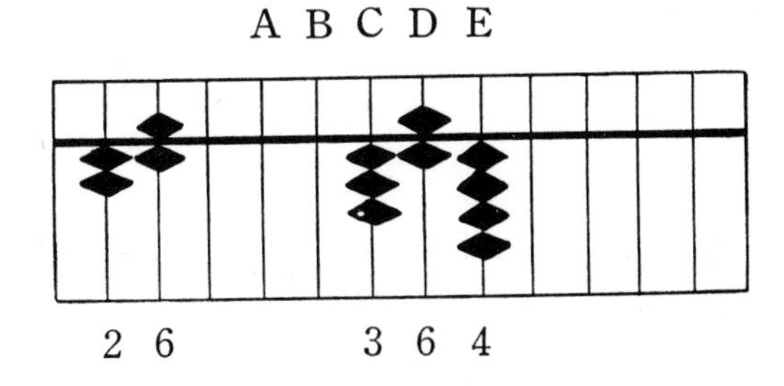

(1) Place the divisor on the left hand side of the board,

and the dividend to the right of it, as is shown in the above illustration.

(2) Beginning with the unit position of the diviend, count off towards the left the number of uprights which are contained in the divisor, and the second upright to the left of that will be the unit position for the quotient.

(3) As the divisor is a two-place number, divide the dividend into groups of two places.

(4) As the first group is 36, we can see that the divisor goes into this only once.

(5) Raise one Earth counter on the second upright to the left of the first upright of the dividend. The board should indicate 10364.

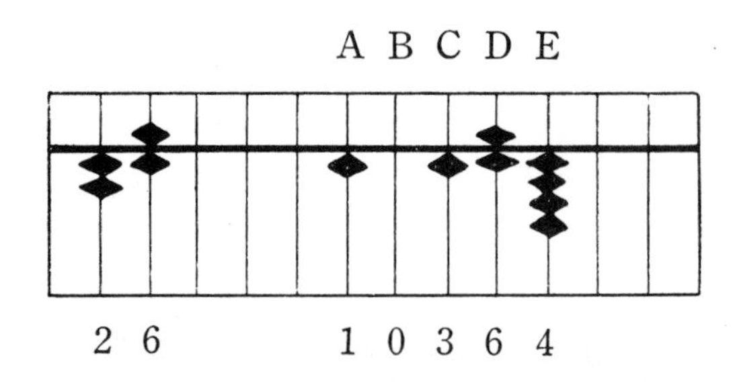

(6) Multiplying the 1 of the temporary quotient by the 2, which is the first figure of the divisor, we get, "1×2 is 2." Return to place two Earth counters on the upright for 100's in the dividend. The board should indicate 10164.

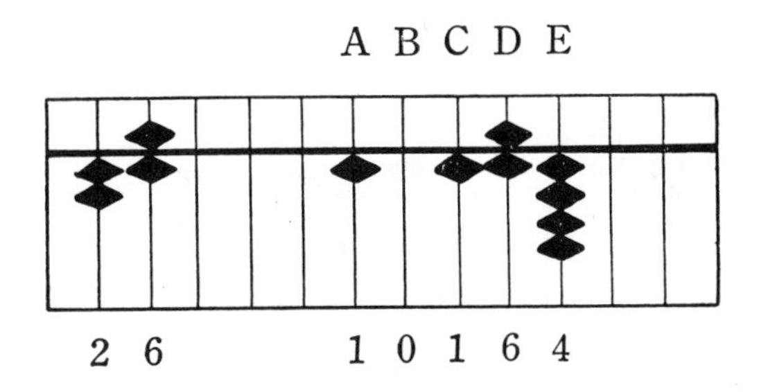

(7) Multiplying the temporary quotient by the second figure of the divisor, we get, "1×6 is 6." Return to place the counters which indicate 6 on the second upright of the first group of the dividend. The board should indicate 10104.

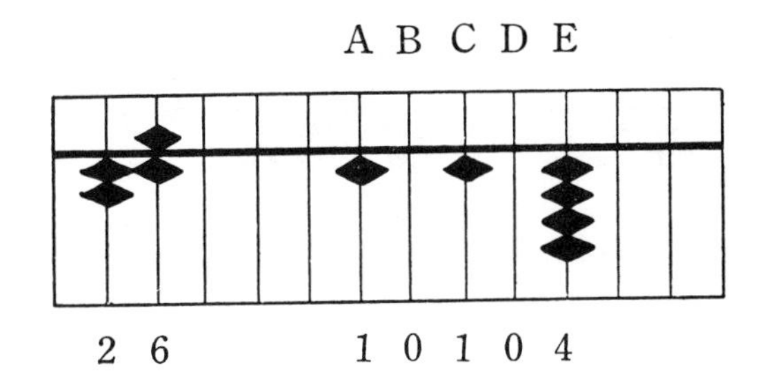

(8) As 26 will not go into 10, we must add one more upright and we find that 26 will go into 104 four times.

(9) We must indicate this 4 on the next upright to the right of the first temporary quotient. The board should now indicate 14104.

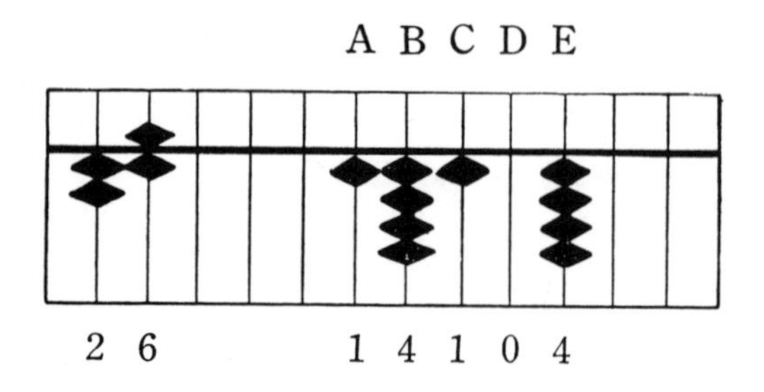

(10) Multiplying the 4 of the second temporary quotient on the upright for 1000's in the dividend by the 2, which is the first figure of the divisor, we get, "2×4 are 8." Return to place 8 of the 10 of the second group. The board should indicate 14024.

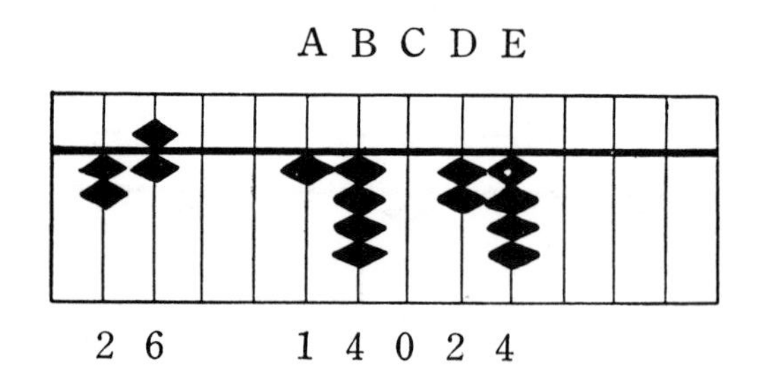

(11) Multiplying the 4 of the second temporary quotient by the 6 which is the second figure of the divisor, we get, "4×6 are 24." Return to place the two Earth counters on the upright for 10's in the dividend and the 4 Earth counters on the last upright.

14 is indicated on the board. It is the quotient.

In the beginning of this example, we found that the unit

position of the quotient was on the upright for 1,000's in the dividend, or the B upright. Therefore, 14 is the real quotient.

By now, I think that division should be plain to you. I hope that you will be diligent in practicing and reviewing, so that you can compute any number easily.

I am glad that by now you can add, subtract, multiply, and divide on the abacus.

Do not be afraid of even the largest numbers, as the process is just the same as you have learned.

Turn to page 193 and compare it with this page.

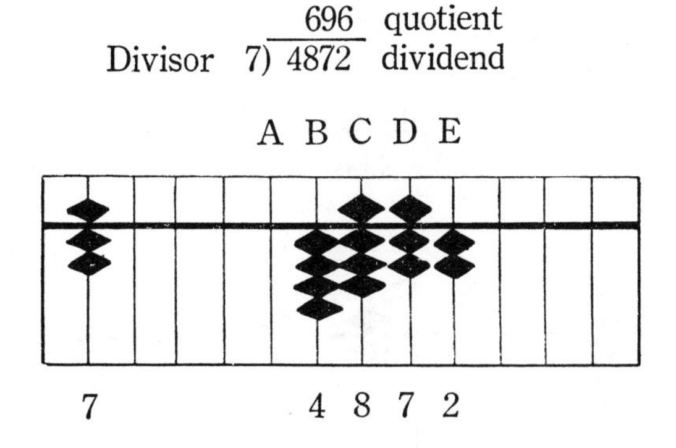

(1) As the divisor is a one upright number, divide the dividend into groups containing one upright each.

(2) As 7 will not go into 4, we must take one more upright into the first group. Then 7 will go into 48 six times.

(3) Place this 6 on the next upright to the left of the first upright of the dividend, or one upright to the right of where it should have gone if we had not borrowed one number from the second group. The board should indicate 64872.

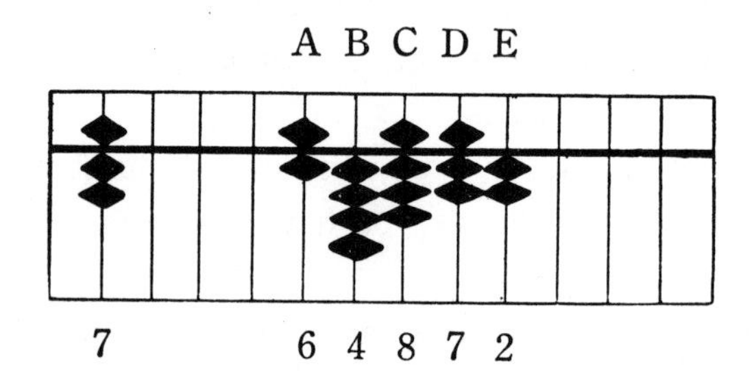

(4) Multiplying the 6 of the first temporary quotient by the divisor, we get, "6×7 are 42." Subtract the 4 from the upright for 1000's in the dividend and the 2 from the upright for 100's in the dividend. The board should indicate 60672.

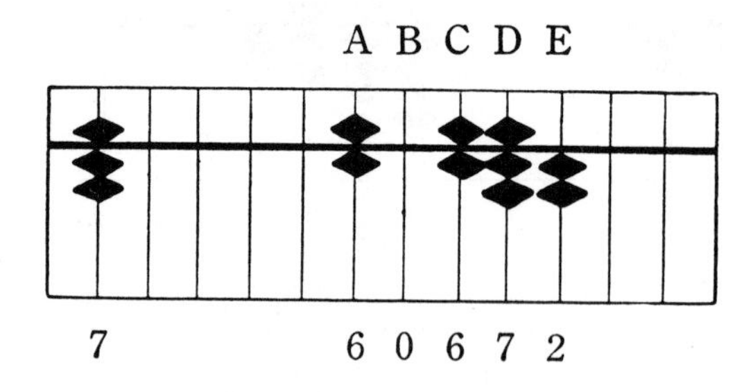

(5) The divisor will not go into the next group, which is 6, therefore, we must borrow one upright. 7 goes into 67 nine times.

(6) Place this product 9 on the next upright to the right of the first temporary quotient, in this case the B upright. The board should indicate 69672.

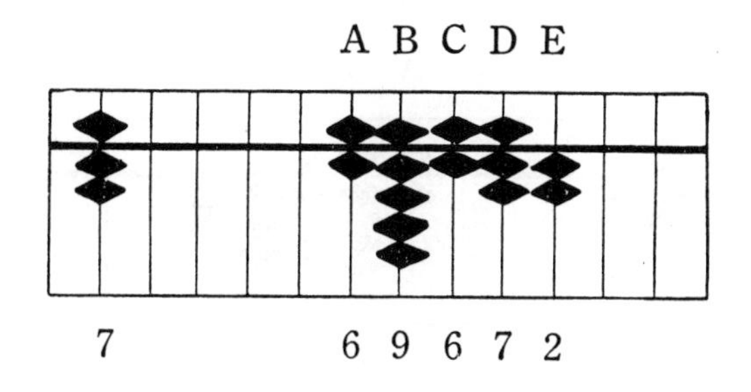

(7) Multiplying the 9 of the second temporary quotient on the upright for 1000's in the dividend by the divisor, we get, "7×9 are 63." Subtract 6 from the upright for 100's in the dividend and 3 from the upright for 10's. The board should indicate 69042.

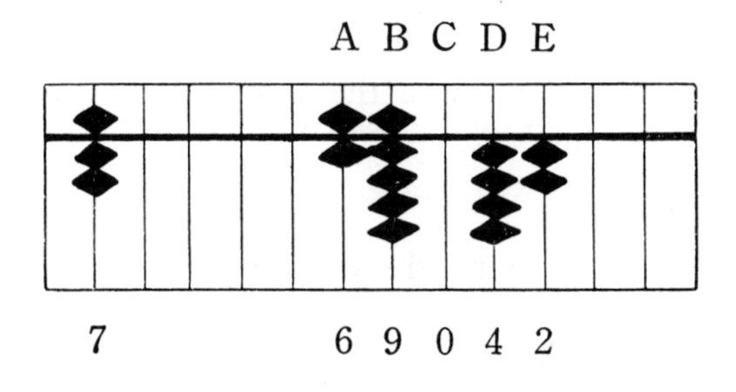

(8) The divisor will not go into the next group, so we must again borrow one upright. Then 7 will go into 42 six times.

(9) Place this 6 on the next upright to the right from the second temporary quotient, in this case the C upright. The board should indicate 69642.

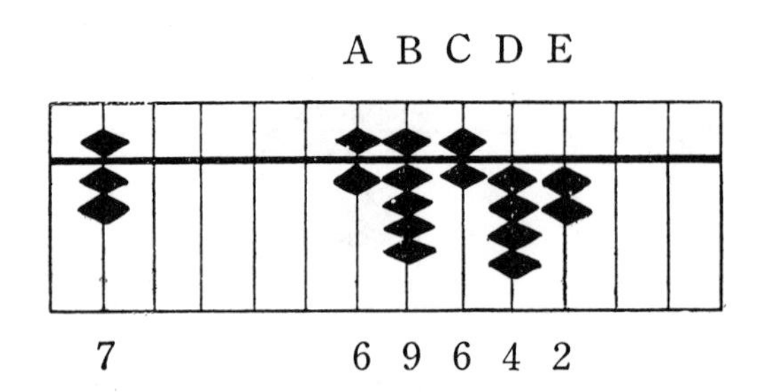

(10) Multiplying the 6 of the third temporary quotient on the upright for 100's in the dividend by the divisor, we get, "6×7 are 42." Subtract the 4 from the upright for 10's in the dividend and the 2 from the last upright.

696 is indicated on the board. It is the quotient.

In the beginning of the division we found that the unit position of the quotient was the C upright. Therefore, 696 is the real quotient,

Find the quotient.

Row A	Row A
(1) 36) 23364	(1) 64) 16512
(2) 43) 31089	(2) 68) 26316

(3) $48\overline{)\,42960}$

(4) $55\overline{)\,50270}$

(5) $59\overline{)\,9558}$

(6) $7\overline{)\,4872}$

(7) $8\overline{)\,6232}$

(8) $9\overline{)\,7758}$

(9) $2\overline{)\,1886}$

(10) $3\overline{)\,468}$

(3) $72\overline{)\,30600}$

(4) $76\overline{)\,39368}$

(5) $85\overline{)\,54315}$

(6) $24\overline{)\,75288}$

(7) $35\overline{)\,186305}$

(8) $41\overline{)\,288558}$

(9) $19\overline{)\,42921}$

(10) $87\overline{)\,64467}$

For the Salesman.

The abacus is very convenient for the salesman in computing change.

The price of the goods is $ 1.26
Clerk receives $ 5.00

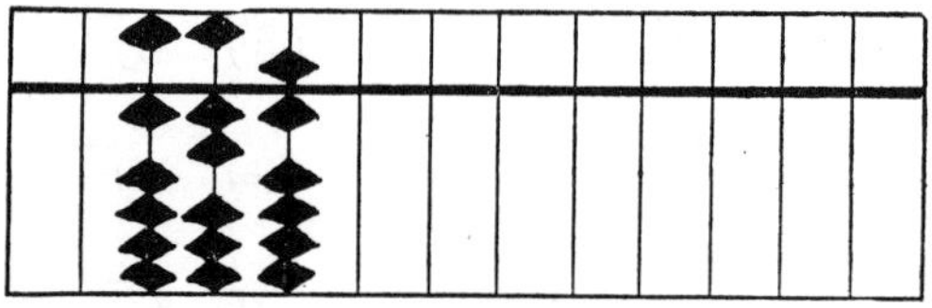

1 2 6 Employed counters
3 7 4 Unemployed counters

(1) As shown in the above illustration the clerk should indicate the price of the goods on the board, in this case it is $ 1.26.

(2) Discounting one on every upright except the last one and the Heaven counter on the first upright, the number of unemployed counters will give the clerk the amount of change required.

(3) 3 counters are unemployed on the upright for 100's, counters indicating 7 on the upright for 10's and 4 counters on the upright for 1's; 374 is indicated. This is the same as $ 3.74, which is the amount of change.

The price of goods $ 3.82
Clerk receives $10.00

3 8 2 Employed counters
6 1 8 Unemployed counters

(1) As is shown in the above illustration, place the price of the goods on the board. $3.82 is indicated as 382.

(2) Find out how many unemployed counters there are, instead of thinking 382 and so many make 1000.

(3) There are 6 on the upright for 100's and 1 on the upright for 10's and 8 on the end upright.

The amount of the change is $6.18.

If the salesman uses an abacus in computing change, he can be not only quick but accurate. Therefore even if the amount is small you should use an abacus.

(1) $ .28¢ Price of goods
$1.00 Money received
$ .72¢ Change

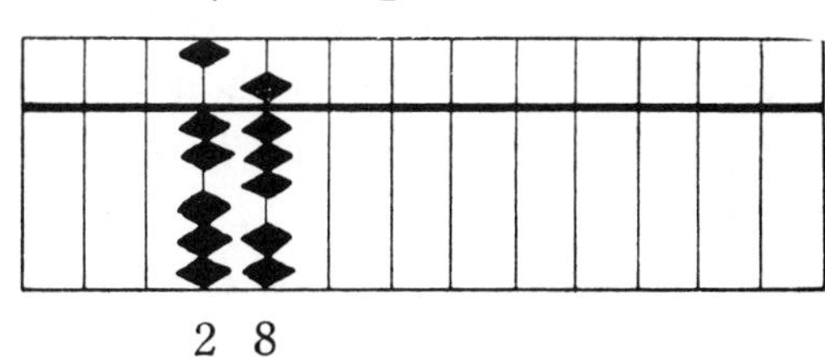

2 8
7 2...............Surplus

(2) $2.73 Price of goods
$5.00 Money received
$2.27 Change

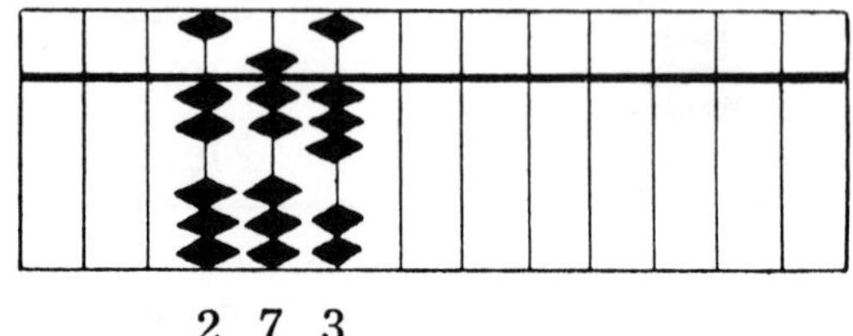

2 7 3
2 2 7..............Surplus

(3) $ 6.19 Price of goods
$10.00 Money received
$ 3.81 Change

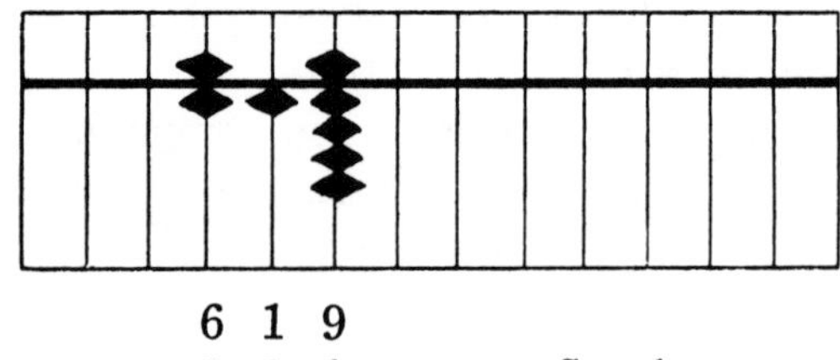

6 1 9
3 8 1..............Surplus

Always when you give back change you should add the amount to the price of the goods to check yourself. If the sum equals the money received you have given back the correct amount.

A CATALOGUE OF SELECTED DOVER BOOKS IN ALL FIELDS OF INTEREST

A CATALOGUE OF SELECTED DOVER BOOKS IN ALL FIELDS OF INTEREST

THE NOTEBOOKS OF LEONARDO DA VINCI, edited by J.P. Richter. Extracts from manuscripts reveal great genius; on painting, sculpture, anatomy, sciences, geography, etc. Both Italian and English. 186 ms. pages reproduced, plus 500 additional drawings, including studies for Last Supper, Sforza monument, etc. 860pp. $7\frac{7}{8} \times 10\frac{3}{4}$. USO 22572-0, 22573-9 Pa., Two vol. set $12.00

ART NOUVEAU DESIGNS IN COLOR, Alphonse Mucha, Maurice Verneuil, Georges Auriol. Full-color reproduction of Combinaisons ornamentales (c. 1900) by Art Nouveau masters. Floral, animal, geometric, interlacings, swashes — borders, frames, spots — all incredibly beautiful. 60 plates, hundreds of designs. $9\frac{3}{8} \times 8\frac{1}{16}$. 22885-1 Pa. $4.00

GRAPHIC WORKS OF ODILON REDON. All great fantastic lithographs, etchings, engravings, drawings, 209 in all. Monsters, Huysmans, still life work, etc. Introduction by Alfred Werner. 209pp. $9\frac{1}{8} \times 12\frac{1}{4}$. 21996-8 Pa. $5.00

EXOTIC FLORAL PATTERNS IN COLOR, E.-A. Seguy. Incredibly beautiful full-color pochoir work by great French designer of 20's. Complete Bouquets et frondaisons, Suggestions pour étoffes. Richness must be seen to be believed. 40 plates containing 120 patterns. 80pp. $9\frac{3}{8} \times 12\frac{1}{4}$. 23041-4 Pa. $6.00

SELECTED ETCHINGS OF JAMES A. MCN. WHISTLER, James A. McN. Whistler. 149 outstanding etchings by the great American artist, including selections from the Thames set and two Venice sets, the complete French set, and many individual prints. Introduction and explanatory note on each print by Maria Naylor. 157pp. $9\frac{3}{8} \times 12\frac{1}{4}$. 23194-1 Pa. $5.00

VISUAL ILLUSIONS: THEIR CAUSES, CHARACTERISTICS, AND APPLICATIONS, Matthew Luckiesh. Thorough description, discussion; shape and size, color, motion; natural illusion. Uses in art and industry. 100 illustrations. 252pp. 21530-X Pa. $2.50

TEN BOOKS ON ARCHITECTURE, Vitruvius. The most important book ever written on architecture. Early Roman aesthetics, technology, classical orders, site selection, all other aspects. Stands behind everything since. Morgan translation. 331pp. 20645-9 Pa. $3.50

THE CODEX NUTTALL, A PICTURE MANUSCRIPT FROM ANCIENT MEXICO, as first edited by Zelia Nuttall. Only inexpensive edition, in full color, of a pre-Columbian Mexican (Mixtec) book. 88 color plates show kings, gods, heroes, temples, sacrifices. New explanatory, historical introduction by Arthur G. Miller. 96pp. $11\frac{3}{8} \times 8\frac{1}{2}$. 23168-2 Pa. $7.50

CREATIVE LITHOGRAPHY AND HOW TO DO IT, Grant Arnold. Lithography as art form: working directly on stone, transfer of drawings, lithotint, mezzotint, color printing; also metal plates. Detailed, thorough. 27 illustrations. 214pp.
21208-4 Pa. $3.00

DESIGN MOTIFS OF ANCIENT MEXICO, Jorge Enciso. Vigorous, powerful ceramic stamp impressions — Maya, Aztec, Toltec, Olmec. Serpents, gods, priests, dancers, etc. 153pp. 6⅛ x 9¼. 20084-1 Pa. $2.50

AMERICAN INDIAN DESIGN AND DECORATION, Leroy Appleton. Full text, plus more than 700 precise drawings of Inca, Maya, Aztec, Pueblo, Plains, NW Coast basketry, sculpture, painting, pottery, sand paintings, metal, etc. 4 plates in color. 279pp. 8⅜ x 11¼. 22704-9 Pa. $4.50

CHINESE LATTICE DESIGNS, Daniel S. Dye. Incredibly beautiful geometric designs: circles, voluted, simple dissections, etc. Inexhaustible source of ideas, motifs. 1239 illustrations. 469pp. 6⅛ x 9¼. 23096-1 Pa. $5.00

JAPANESE DESIGN MOTIFS, Matsuya Co. Mon, or heraldic designs. Over 4000 typical, beautiful designs: birds, animals, flowers, swords, fans, geometric; all beautifully stylized. 213pp. 11⅜ x 8¼. 22874-6 Pa. $4.95

PERSPECTIVE, Jan Vredeman de Vries. 73 perspective plates from 1604 edition; buildings, townscapes, stairways, fantastic scenes. Remarkable for beauty, surrealistic atmosphere; real eye-catchers. Introduction by Adolf Placzek. 74pp. 11⅜ x 8¼. 20186-4 Pa. $2.75

EARLY AMERICAN DESIGN MOTIFS, Suzanne E. Chapman. 497 motifs, designs, from painting on wood, ceramics, appliqué, glassware, samplers, metal work, etc. Florals, landscapes, birds and animals, geometrics, letters, etc. Inexhaustible. Enlarged edition. 138pp. 8⅜ x 11¼. 22985-8 Pa. $3.50
23084-8 Clothbd. $7.95

VICTORIAN STENCILS FOR DESIGN AND DECORATION, edited by E.V. Gillon, Jr. 113 wonderful ornate Victorian pieces from German sources; florals, geometrics; borders, corner pieces; bird motifs, etc. 64pp. 9⅜ x 12¼. 21995-X Pa. $2.50

ART NOUVEAU: AN ANTHOLOGY OF DESIGN AND ILLUSTRATION FROM THE STUDIO, edited by E.V. Gillon, Jr. Graphic arts: book jackets, posters, engravings, illustrations, decorations; Crane, Beardsley, Bradley and many others. Inexhaustible. 92pp. 8⅛ x 11. 22388-4 Pa. $2.50

ORIGINAL ART DECO DESIGNS, William Rowe. First-rate, highly imaginative modern Art Deco frames, borders, compositions, alphabets, florals, insectals, Wurlitzer-types, etc. Much finest modern Art Deco. 80 plates, 8 in color. 8⅜ x 11¼. 22567-4 Pa. $3.00

HANDBOOK OF DESIGNS AND DEVICES, Clarence P. Hornung. Over 1800 basic geometric designs based on circle, triangle, square, scroll, cross, etc. Largest such collection in existence. 261pp. 20125-2 Pa. $2.50

150 MASTERPIECES OF DRAWING, edited by Anthony Toney. 150 plates, early 15th century to end of 18th century; Rembrandt, Michelangelo, Dürer, Fragonard, Watteau, Wouwerman, many others. 150pp. 8³/8 x 11¼. 21032-4 Pa. $3.50

THE GOLDEN AGE OF THE POSTER, Hayward and Blanche Cirker. 70 extraordinary posters in full colors, from Maîtres de l'Affiche, Mucha, Lautrec, Bradley, Cheret, Beardsley, many others. 9³/8 x 12¼. 22753-7 Pa. $4.95
21718-3 Clothbd. $7.95

SIMPLICISSIMUS, selection, translations and text by Stanley Appelbaum. 180 satirical drawings, 16 in full color, from the famous German weekly magazine in the years 1896 to 1926. 24 artists included: Grosz, Kley, Pascin, Kubin, Kollwitz, plus Heine, Thöny, Bruno Paul, others. 172pp. 8½ x 12¼. 23098-8 Pa. $5.00
23099-6 Clothbd. $10.00

THE EARLY WORK OF AUBREY BEARDSLEY, Aubrey Beardsley. 157 plates, 2 in color: Manon Lescaut, Madame Bovary, Morte d'Arthur, Salome, other. Introduction by H. Marillier. 175pp. 8½ x 11. 21816-3 Pa. $3.50

THE LATER WORK OF AUBREY BEARDSLEY, Aubrey Beardsley. Exotic masterpieces of full maturity: Venus and Tannhäuser, Lysistrata, Rape of the Lock, Volpone, Savoy material, etc. 174 plates, 2 in color. 176pp. 8½ x 11. 21817-1 Pa. $3.75

DRAWINGS OF WILLIAM BLAKE, William Blake. 92 plates from Book of Job, Divine Comedy, Paradise Lost, visionary heads, mythological figures, Laocoön, etc. Selection, introduction, commentary by Sir Geoffrey Keynes. 178pp. 8½ x 11.
22303-5 Pa. $3.50

LONDON: A PILGRIMAGE, Gustave Doré, Blanchard Jerrold. Squalor, riches, misery, beauty of mid-Victorian metropolis; 55 wonderful plates, 125 other illustrations, full social, cultural text by Jerrold. 191pp. of text. 8¹/8 x 11.
22306-X Pa. $5.00

THE COMPLETE WOODCUTS OF ALBRECHT DÜRER, edited by Dr. W. Kurth. 346 in all: Old Testament, St. Jerome, Passion, Life of Virgin, Apocalypse, many others. Introduction by Campbell Dodgson. 285pp. 8½ x 12¼. 21097-9 Pa. $6.00

THE DISASTERS OF WAR, Francisco Goya. 83 etchings record horrors of Napoleonic wars in Spain and war in general. Reprint of 1st edition, plus 3 additional plates. Introduction by Philip Hofer. 97pp. 9³/8 x 8¼. 21872-4 Pa. $2.50

ENGRAVINGS OF HOGARTH, William Hogarth. 101 of Hogarth's greatest works: Rake's Progress, Harlot's Progress, Illustrations for Hudibras, Midnight Modern Conversation, Before and After, Beer Street and Gin Lane, many more. Full commentary. 256pp. 11 x 14. 22479-1 Pa. $6.00
23023-6 Clothbd. $13.50

PRIMITIVE ART, Franz Boas. Great anthropologist on ceramics, textiles, wood, stone, metal, etc.; patterns, technology, symbols, styles. All areas, but fullest on Northwest Coast Indians. 350 illustrations. 378pp. 20025-6 Pa. $3.50

EARLY NEW ENGLAND GRAVESTONE RUBBINGS, Edmund V. Gillon, Jr. 43 photographs, 226 rubbings show heavily symbolic, macabre, sometimes humorous primitive American art. Up to early 19th century. 207pp. 8³/8 x 11¼.
21380-3 Pa. $4.00

L.J.M. DAGUERRE: THE HISTORY OF THE DIORAMA AND THE DAGUERREOTYPE, Helmut and Alison Gernsheim. Definitive account. Early history, life and work of Daguerre; discovery of daguerreotype process; diffusion abroad; other early photography. 124 illustrations. 226pp. 6¹/6 x 9¼. 22290-X Pa. $4.00

PHOTOGRAPHY AND THE AMERICAN SCENE, Robert Taft. The basic book on American photography as art, recording form, 1839-1889. Development, influence on society, great photographers, types (portraits, war, frontier, etc.), whatever else needed. Inexhaustible. Illustrated with 322 early photos, daguerreotypes, tintypes, stereo slides, etc. 546pp. 6¹/8 x 9¼. 21201-7 Pa. $5.00

PHOTOGRAPHIC SKETCHBOOK OF THE CIVIL WAR, Alexander Gardner. Reproduction of 1866 volume with 100 on-the-field photographs: Manassas, Lincoln on battlefield, slave pens, etc. Introduction by E.F. Bleiler. 224pp. 10¾ x 9.
22731-6 Pa. $4.50

THE MOVIES: A PICTURE QUIZ BOOK, Stanley Appelbaum & Hayward Cirker. Match stars with their movies, name actors and actresses, test your movie skill with 241 stills from 236 great movies, 1902-1959. Indexes of performers and films. 128pp. 8³/8 x 9¼. 20222-4 Pa. $2.50

THE TALKIES, Richard Griffith. Anthology of features, articles from Photoplay, 1928-1940, reproduced complete. Stars, famous movies, technical features, fabulous ads, etc.; Garbo, Chaplin, King Kong, Lubitsch, etc. 4 color plates, scores of illustrations. 327pp. 8³/8 x 11¼. 22762-6 Pa. $5.95

THE MOVIE MUSICAL FROM VITAPHONE TO "42ND STREET," edited by Miles Kreuger. Relive the rise of the movie musical as reported in the pages of Photoplay magazine (1926-1933): every movie review, cast list, ad, and record review; every significant feature article, production still, biography, forecast, and gossip story. Profusely illustrated. 367pp. 8³/8 x 11¼. 23154-2 Pa. $6.95

JOHANN SEBASTIAN BACH, Philipp Spitta. Great classic of biography, musical commentary, with hundreds of pieces analyzed. Also good for Bach's contemporaries. 450 musical examples. Total of 1799pp.
EUK 22278-0, 22279-9 Clothbd., Two vol. set $25.00

BEETHOVEN AND HIS NINE SYMPHONIES, Sir George Grove. Thorough history, analysis, commentary on symphonies and some related pieces. For either beginner or advanced student. 436 musical passages. 407pp. 20334-4 Pa. $4.00

MOZART AND HIS PIANO CONCERTOS, Cuthbert Girdlestone. The only full-length study. Detailed analyses of all 21 concertos, sources; 417 musical examples. 509pp. 21271-8 Pa. $4.50

THE FITZWILLIAM VIRGINAL BOOK, edited by J. Fuller Maitland, W.B. Squire. Famous early 17th century collection of keyboard music, 300 works by Morley, Byrd, Bull, Gibbons, etc. Modern notation. Total of 938pp. 8$^{3}/_{8}$ x 11.
ECE 21068-5, 21069-3 Pa., Two vol. set $12.00

COMPLETE STRING QUARTETS, Wolfgang A. Mozart. Breitkopf and Härtel edition. All 23 string quartets plus alternate slow movement to K156. Study score. 277pp. 9$^{3}/_{8}$ x 12$^{1}/_{4}$. 22372-8 Pa. $6.00

COMPLETE SONG CYCLES, Franz Schubert. Complete piano, vocal music of Die Schöne Müllerin, Die Winterreise, Schwanengesang. Also Drinker English singing translations. Breitkopf and Härtel edition. 217pp. 9$^{3}/_{8}$ x 12$^{1}/_{4}$.
22649-2 Pa. $4.00

THE COMPLETE PRELUDES AND ETUDES FOR PIANOFORTE SOLO, Alexander Scriabin. All the preludes and etudes including many perfectly spun miniatures. Edited by K.N. Igumnov and Y.I. Mil'shteyn. 250pp. 9 x 12. 22919-X Pa. $5.00

TRISTAN UND ISOLDE, Richard Wagner. Full orchestral score with complete instrumentation. Do not confuse with piano reduction. Commentary by Felix Mottl, great Wagnerian conductor and scholar. Study score. 655pp. 8$^{1}/_{8}$ x 11.
22915-7 Pa. $10.00

FAVORITE SONGS OF THE NINETIES, ed. Robert Fremont. Full reproduction, including covers, of 88 favorites: Ta-Ra-Ra-Boom-De-Aye, The Band Played On, Bird in a Gilded Cage, Under the Bamboo Tree, After the Ball, etc. 401pp. 9 x 12.
EBE 21536-9 Pa. $6.95

SOUSA'S GREAT MARCHES IN PIANO TRANSCRIPTION: ORIGINAL SHEET MUSIC OF 23 WORKS, John Philip Sousa. Selected by Lester S. Levy. Playing edition includes: The Stars and Stripes Forever, The Thunderer, The Gladiator, King Cotton, Washington Post, much more. 24 illustrations. 111pp. 9 x 12.
USO 23132-1 Pa. $3.50

CLASSIC PIANO RAGS, selected with an introduction by Rudi Blesh. Best ragtime music (1897-1922) by Scott Joplin, James Scott, Joseph F. Lamb, Tom Turpin, 9 others. Printed from best original sheet music, plus covers. 364pp. 9 x 12.
EBE 20469-3 Pa. $6.95

ANALYSIS OF CHINESE CHARACTERS, C.D. Wilder, J.H. Ingram. 1000 most important characters analyzed according to primitives, phonetics, historical development. Traditional method offers mnemonic aid to beginner, intermediate student of Chinese, Japanese. 365pp. 23045-7 Pa. $4.00

MODERN CHINESE: A BASIC COURSE, Faculty of Peking University. Self study, classroom course in modern Mandarin. Records contain phonetics, vocabulary, sentences, lessons. 249 page book contains all recorded text, translations, grammar, vocabulary, exercises. Best course on market. 3 12" 33$^{1}/_{3}$ monaural records, book, album. 98832-5 Set $12.50

THE BEST DR. THORNDYKE DETECTIVE STORIES, R. Austin Freeman. The Case of Oscar Brodski, The Moabite Cipher, and 5 other favorites featuring the great scientific detective, plus his long-believed-lost first adventure — 31 New Inn — reprinted here for the first time. Edited by E.F. Bleiler. USO 20388-3 Pa. $3.00

BEST "THINKING MACHINE" DETECTIVE STORIES, Jacques Futrelle. The Problem of Cell 13 and 11 other stories about Prof. Augustus S.F.X. Van Dusen, including two "lost" stories. First reprinting of several. Edited by E.F. Bleiler. 241pp. 20537-1 Pa. $3.00

UNCLE SILAS, J. Sheridan LeFanu. Victorian Gothic mystery novel, considered by many best of period, even better than Collins or Dickens. Wonderful psychological terror. Introduction by Frederick Shroyer. 436pp. 21715-9 Pa. $4.00

BEST DR. POGGIOLI DETECTIVE STORIES, T.S. Stribling. 15 best stories from EQMM and The Saint offer new adventures in Mexico, Florida, Tennessee hills as Poggioli unravels mysteries and combats Count Jalacki. 217pp. 23227-1 Pa. $3.00

EIGHT DIME NOVELS, selected with an introduction by E.F. Bleiler. Adventures of Old King Brady, Frank James, Nick Carter, Deadwood Dick, Buffalo Bill, The Steam Man, Frank Merriwell, and Horatio Alger — 1877 to 1905. Important, entertaining popular literature in facsimile reprint, with original covers. 190pp. 9 x 12. 22975-0 Pa. $3.50

ALICE'S ADVENTURES UNDER GROUND, Lewis Carroll. Facsimile of ms. Carroll gave Alice Liddell in 1864. Different in many ways from final Alice. Handlettered, illustrated by Carroll. Introduction by Martin Gardner. 128pp. 21482-6 Pa. $1.50

ALICE IN WONDERLAND COLORING BOOK, Lewis Carroll. Pictures by John Tenniel. Large-size versions of the famous illustrations of Alice, Cheshire Cat, Mad Hatter and all the others, waiting for your crayons. Abridged text. 36 illustrations. 64pp. 8¼ x 11. 22853-3 Pa. $1.50

AVENTURES D'ALICE AU PAYS DES MERVEILLES, Lewis Carroll. Bué's translation of "Alice" into French, supervised by Carroll himself. Novel way to learn language. (No English text.) 42 Tenniel illustrations. 196pp. 22836-3 Pa. $2.00

MYTHS AND FOLK TALES OF IRELAND, Jeremiah Curtin. 11 stories that are Irish versions of European fairy tales and 9 stories from the Fenian cycle — 20 tales of legend and magic that comprise an essential work in the history of folklore. 256pp. 22430-9 Pa. $3.00

EAST O' THE SUN AND WEST O' THE MOON, George W. Dasent. Only full edition of favorite, wonderful Norwegian fairytales — Why the Sea is Salt, Boots and the Troll, etc. — with 77 illustrations by Kittelsen & Werenskiöld. 418pp. 22521-6 Pa. $3.50

PERRAULT'S FAIRY TALES, Charles Perrault and Gustave Doré. Original versions of Cinderella, Sleeping Beauty, Little Red Riding Hood, etc. in best translation, with 34 wonderful illustrations by Gustave Doré. 117pp. 8⅛ x 11. 22311-6 Pa. $2.50

MOTHER GOOSE'S MELODIES. Facsimile of fabulously rare Munroe and Francis "copyright 1833" Boston edition. Familiar and unusual rhymes, wonderful old woodcut illustrations. Edited by E.F. Bleiler. 128pp. 4½ x 6⅜. 22577-1 Pa. $1.00

MOTHER GOOSE IN HIEROGLYPHICS. Favorite nursery rhymes presented in rebus form for children. Fascinating 1849 edition reproduced in toto, with key. Introduction by E.F. Bleiler. About 400 woodcuts. 64pp. 6⅞ x 5¼. 20745-5 Pa. $1.00

PETER PIPER'S PRACTICAL PRINCIPLES OF PLAIN & PERFECT PRONUNCIATION. Alliterative jingles and tongue-twisters. Reproduction in full of 1830 first American edition. 25 spirited woodcuts. 32pp. 4½ x 6⅜. 22560-7 Pa. $1.00

MARMADUKE MULTIPLY'S MERRY METHOD OF MAKING MINOR MATHEMATICIANS. Fellow to Peter Piper, it teaches multiplication table by catchy rhymes and woodcuts. 1841 Munroe & Francis edition. Edited by E.F. Bleiler. 103pp. 4⅝ x 6.
22773-1 Pa. $1.25
20171-6 Clothbd. $3.00

THE NIGHT BEFORE CHRISTMAS, Clement Moore. Full text, and woodcuts from original 1848 book. Also critical, historical material. 19 illustrations. 40pp. 4⅝ x 6. 22797-9 Pa. $1.00

THE KING OF THE GOLDEN RIVER, John Ruskin. Victorian children's classic of three brothers, their attempts to reach the Golden River, what becomes of them. Facsimile of original 1889 edition. 22 illustrations. 56pp. 4⅝ x 6⅜.
20066-3 Pa. $1.25

DREAMS OF THE RAREBIT FIEND, Winsor McCay. Pioneer cartoon strip, unexcelled for beauty, imagination, in 60 full sequences. Incredible technical virtuosity, wonderful visual wit. Historical introduction. 62pp. 8⅜ x 11¼. 21347-1 Pa. $2.00

THE KATZENJAMMER KIDS, Rudolf Dirks. In full color, 14 strips from 1906-7; full of imagination, characteristic humor. Classic of great historical importance. Introduction by August Derleth. 32pp. 9¼ x 12¼. 23005-8 Pa. $2.00

LITTLE ORPHAN ANNIE AND LITTLE ORPHAN ANNIE IN COSMIC CITY, Harold Gray. Two great sequences from the early strips: our curly-haired heroine defends the Warbucks' financial empire and, then, takes on meanie Phineas P. Pinchpenny. Leapin' lizards! 178pp. 6⅛ x 8⅜. 23107-0 Pa. $2.00

WHEN A FELLER NEEDS A FRIEND, Clare Briggs. 122 cartoons by one of the greatest newspaper cartoonists of the early 20th century — about growing up, making a living, family life, daily frustrations and occasional triumphs. 121pp. 8½ x 9½.
23148-8 Pa. $2.50

THE BEST OF GLUYAS WILLIAMS. 100 drawings by one of America's finest cartoonists: The Day a Cake of Ivory Soap Sank at Proctor & Gamble's, At the Life Insurance Agents' Banquet, and many other gems from the 20's and 30's. 118pp. 8⅜ x 11¼. 22737-5 Pa. $2.50

THE MAGIC MOVING PICTURE BOOK, Bliss, Sands & Co. The pictures in this book move! Volcanoes erupt, a house burns, a serpentine dancer wiggles her way through a number. By using a specially ruled acetate screen provided, you can obtain these and 15 other startling effects. Originally "The Motograph Moving Picture Book." 32pp. 8¼ x 11. 23224-7 Pa. $1.75

STRING FIGURES AND HOW TO MAKE THEM, Caroline F. Jayne. Fullest, clearest instructions on string figures from around world: Eskimo, Navajo, Lapp, Europe, more. Cats cradle, moving spear, lightning, stars. Introduction by A.C. Haddon. 950 illustrations. 407pp. 20152-X Pa. $3.00

PAPER FOLDING FOR BEGINNERS, William D. Murray and Francis J. Rigney. Clearest book on market for making origami sail boats, roosters, frogs that move legs, cups, bonbon boxes. 40 projects. More than 275 illustrations. Photographs. 94pp. 20713-7 Pa. $1.25

INDIAN SIGN LANGUAGE, William Tomkins. Over 525 signs developed by Sioux, Blackfoot, Cheyenne, Arapahoe and other tribes. Written instructions and diagrams: how to make words, construct sentences. Also 290 pictographs of Sioux and Ojibway tribes. 111pp. 6⅛ x 9¼. 22029-X Pa. $1.50

BOOMERANGS: HOW TO MAKE AND THROW THEM, Bernard S. Mason. Easy to make and throw, dozens of designs: cross-stick, pinwheel, boomabird, tumblestick, Australian curved stick boomerang. Complete throwing instructions. All safe. 99pp. 23028-7 Pa. $1.50

25 KITES THAT FLY, Leslie Hunt. Full, easy to follow instructions for kites made from inexpensive materials. Many novelties. Reeling, raising, designing your own. 70 illustrations. 110pp. 22550-X Pa. $1.25

TRICKS AND GAMES ON THE POOL TABLE, Fred Herrmann. 79 tricks and games, some solitaires, some for 2 or more players, some competitive; mystifying shots and throws, unusual carom, tricks involving cork, coins, a hat, more. 77 figures. 95pp. 21814-7 Pa. $1.25

WOODCRAFT AND CAMPING, Bernard S. Mason. How to make a quick emergency shelter, select woods that will burn immediately, make do with limited supplies, etc. Also making many things out of wood, rawhide, bark, at camp. Formerly titled Woodcraft. 295 illustrations. 580pp. 21951-8 Pa. $4.00

AN INTRODUCTION TO CHESS MOVES AND TACTICS SIMPLY EXPLAINED, Leonard Barden. Informal intermediate introduction: reasons for moves, tactics, openings, traps, positional play, endgame. Isolates patterns. 102pp. USO 21210-6 Pa. $1.35

LASKER'S MANUAL OF CHESS, Dr. Emanuel Lasker. Great world champion offers very thorough coverage of all aspects of chess. Combinations, position play, openings, endgame, aesthetics of chess, philosophy of struggle, much more. Filled with analyzed games. 390pp. 20640-8 Pa. $3.50

DRIED FLOWERS, Sarah Whitlock and Martha Rankin. Concise, clear, practical guide to dehydration, glycerinizing, pressing plant material, and more. Covers use of silica gel. 12 drawings. Originally titled "New Techniques with Dried Flowers." 32pp. 21802-3 Pa. $1.00

ABC OF POULTRY RAISING, J.H. Florea. Poultry expert, editor tells how to raise chickens on home or small business basis. Breeds, feeding, housing, laying, etc. Very concrete, practical. 50 illustrations. 256pp. 23201-8 Pa. $3.00

HOW INDIANS USE WILD PLANTS FOR FOOD, MEDICINE & CRAFTS, Frances Densmore. Smithsonian, Bureau of American Ethnology report presents wealth of material on nearly 200 plants used by Chippewas of Minnesota and Wisconsin. 33 plates plus 122pp. of text. 6⅛ x 9¼. 23019-8 Pa. $2.50

THE HERBAL OR GENERAL HISTORY OF PLANTS, John Gerard. The 1633 edition revised and enlarged by Thomas Johnson. Containing almost 2850 plant descriptions and 2705 superb illustrations, Gerard's Herbal is a monumental work, the book all modern English herbals are derived from, and the one herbal every serious enthusiast should have in its entirety. Original editions are worth perhaps $750. 1678pp. 8½ x 12¼. 23147-X Clothbd. $50.00

A MODERN HERBAL, Margaret Grieve. Much the fullest, most exact, most useful compilation of herbal material. Gigantic alphabetical encyclopedia, from aconite to zedoary, gives botanical information, medical properties, folklore, economic uses, and much else. Indispensable to serious reader. 161 illustrations. 888pp. 6½ x 9¼. USO 22798-7, 22799-5 Pa., Two vol. set $10.00

HOW TO KNOW THE FERNS, Frances T. Parsons. Delightful classic. Identification, fern lore, for Eastern and Central U.S.A. Has introduced thousands to interesting life form. 99 illustrations. 215pp. 20740-4 Pa. $2.50

THE MUSHROOM HANDBOOK, Louis C.C. Krieger. Still the best popular handbook. Full descriptions of 259 species, extremely thorough text, habitats, luminescence, poisons, folklore, etc. 32 color plates; 126 other illustrations. 560pp. 21861-9 Pa. $4.50

HOW TO KNOW THE WILD FRUITS, Maude G. Peterson. Classic guide covers nearly 200 trees, shrubs, smaller plants of the U.S. arranged by color of fruit and then by family. Full text provides names, descriptions, edibility, uses. 80 illustrations. 400pp. 22943-2 Pa. $3.00

COMMON WEEDS OF THE UNITED STATES, U.S. Department of Agriculture. Covers 220 important weeds with illustration, maps, botanical information, plant lore for each. Over 225 illustrations. 463pp. 6⅛ x 9¼. 20504-5 Pa. $4.50

HOW TO KNOW THE WILD FLOWERS, Mrs. William S. Dana. Still best popular book for East and Central USA. Over 500 plants easily identified, with plant lore; arranged according to color and flowering time. 174 plates. 459pp. 20332-8 Pa. $3.50

HOUDINI ON MAGIC, Harold Houdini. Edited by Walter Gibson, Morris N. Young. How he escaped; exposés of fake spiritualists; instructions for eye-catching tricks; other fascinating material by and about greatest magician. 155 illustrations. 280pp. 20384-0 Pa. $2.50

HANDBOOK OF THE NUTRITIONAL CONTENTS OF FOOD, U.S. Dept. of Agriculture. Largest, most detailed source of food nutrition information ever prepared. Two mammoth tables: one measuring nutrients in 100 grams of edible portion; the other, in edible portion of 1 pound as purchased. Originally titled Composition of Foods. 190pp. 9 x 12. 21342-0 Pa. $4.00

COMPLETE GUIDE TO HOME CANNING, PRESERVING AND FREEZING, U.S. Dept. of Agriculture. Seven basic manuals with full instructions for jams and jellies; pickles and relishes; canning fruits, vegetables, meat; freezing anything. Really good recipes, exact instructions for optimal results. Save a fortune in food. 156 illustrations. 214pp. 6¹/₈ x 9¼. 22911-4 Pa. $2.50

THE BREAD TRAY, Louis P. De Gouy. Nearly every bread the cook could buy or make: bread sticks of Italy, fruit breads of Greece, glazed rolls of Vienna, everything from corn pone to croissants. Over 500 recipes altogether. including buns, rolls, muffins, scones, and more. 463pp. 23000-7 Pa. $3.50

CREATIVE HAMBURGER COOKERY, Louis P. De Gouy. 182 unusual recipes for casseroles, meat loaves and hamburgers that turn inexpensive ground meat into memorable main dishes: Arizona chili burgers, burger tamale pie, burger stew, burger corn loaf, burger wine loaf, and more. 120pp. 23001-5 Pa. $1.75

LONG ISLAND SEAFOOD COOKBOOK, J. George Frederick and Jean Joyce. Probably the best American seafood cookbook. Hundreds of recipes. 40 gourmet sauces, 123 recipes using oysters alone! All varieties of fish and seafood amply represented. 324pp. 22677-8 Pa. $3.00

THE EPICUREAN: A COMPLETE TREATISE OF ANALYTICAL AND PRACTICAL STUDIES IN THE CULINARY ART, Charles Ranhofer. Great modern classic. 3,500 recipes from master chef of Delmonico's, turn-of-the-century America's best restaurant. Also explained, many techniques known only to professional chefs. 775 illustrations. 1183pp. 6⁵/₈ x 10. 22680-8 Clothbd. $17.50

THE AMERICAN WINE COOK BOOK, Ted Hatch. Over 700 recipes: old favorites livened up with wine plus many more: Czech fish soup, quince soup, sauce Perigueux, shrimp shortcake, filets Stroganoff, cordon bleu goulash, jambonneau, wine fruit cake, more. 314pp. 22796-0 Pa. $2.50

DELICIOUS VEGETARIAN COOKING, Ivan Baker. Close to 500 delicious and varied recipes: soups, main course dishes (pea, bean, lentil, çheese, vegetable, pasta, and egg dishes), savories, stews, whole-wheat breads and cakes, more. 168pp. USO 22834-7 Pa. $1.75